Salmonella - Current Trends and Perspectives in Detection and Control

Edited by Chenxi Huang

Published in London, United Kingdom

Salmonella - Current Trends and Perspectives in Detection and Control
http://dx.doi.org/10.5772/intechopen.1001722
Edited by Chenxi Huang

Contributors
Abdul Kabir, Amina Tufail, Anees Ur Rahman, Burak Sayar, Chenxi Huang, Das Shukla, Emine Kübra Dindar Demiray, Momin Khan, Nirmal Kirti, Rui Liu, Singha S. Krishna, Yusra Ashfaq

First published in London, United Kingdom, 2024 by IntechOpen
IntechOpen is the global imprint of INTECHOPEN LIMITED, registered in England and Wales, registration number: 11086078, 167-169 Great Portland Street, London, W1W 5PF, United Kingdom

British Library Cataloguing-in-Publication Data
A catalogue record for this book is available from the British Library

Additional hard and PDF copies can be obtained from orders@intechopen.com

Salmonella - Current Trends and Perspectives in Detection and Control
Edited by Chenxi Huang
p. cm.
Print ISBN 978-0-85466-707-9
Online ISBN 978-0-85466-706-2
eBook (PDF) ISBN 978-0-85466-708-6

For EU product safety concerns:
IN TECH d.o.o., Prolaz Marije Krucifikse Kozulić 3, 51000 Rijeka, Croatia,
info@intechopen.com or visit our website at intechopen.com.

Meet the editor

Chenxi Huang is a full-time assistant professor at the School of Life Sciences, Hubei University, China. She obtained a Ph.D. from Huazhong Agricultural University, China. In the last five years, she has contributed as co-author to thirty-four peer-reviewed papers and two book chapters on foodborne pathogens. She is an active reviewer for various food science and microbiology journals and is an editorial board member for two international journals. Her research focuses on biosensors, molecular technology, and phage-based technology to support the development of novel, highly specific, rapid, low-cost methods for quality control as well as separating and detecting Salmonella and other foodborne pathogens and contaminants.

Contents

Preface

It is my great honor to have the opportunity to guide readers in this important area related to food safety and global health. In this book, we focus on *Salmonella* and current trends and perspectives in its detection and control. With its complete summary and cutting-edge insights, this book reveals the impact of salmonellosis at different levels and its response strategies.

In Chapter 1, Emine Kübra Dindar Demiray and Burak Sayar not only outline the epidemiological characteristics of salmonellosis but also discuss the practical strategies and potentials for preventing and controlling *Salmonella* infections. They then discuss the challenges and management of *Salmonella* disease.

In Chapter 2, Amina Tufail and Yusra Ashfaq provide an in-depth analysis of birds as important carriers of *Salmonella*. It explores the ecological role of birds in salmonellosis epidemics and analyzes the mechanisms of salmonellosis transmission dynamics through epidemiological evidence, genomic analysis, and case studies. The ecological roles and transmission dynamics of *Salmonella* provide a basis for understanding and intervention of salmonellosis.

In Chapter 3, Abdul Kabir, Momin Khan and Anees Ur Rahman focus on *Salmonella* identification in calves. It summarizes the advanced methods and techniques, including molecular biology methods and biosensors, that can achieve accurate identification of *Salmonella* in calves, providing reliable and accurate information on *Salmonella* enterica serotypes.

In Chapter 4, Kirti et al. review techniques ranging from conventional methods to novel non-culture methods like molecular-based, immunological-based, mass spectrometry, optical phenotyping, and biosensor methods. They also discuss the prevention and control of antimicrobial treatment and drug resistance against *Salmonella* infections.

Finally, in Chapter 5, Rui Liu and Chenxi Huang illustrate phages as recognition elements and antibacterial agents for rapid detection and biocontrol of *Salmonella* in foodstuffs and agriculture-related matrices. Various analysis and biocontrol technologies for *Salmonella* are carefully compared by their advantages and disadvantages to guide further development of more efficient analytical methods.

To sum up, this book is a useful reference for the detection and prevention of *Salmonella* infection. It is not only a resource for researchers in the fields of food safety and microbiology but also a guide for policymakers and industry managers. I am enormously grateful to all the authors and the staff at IntechOpen who have contributed to making this book a reality. I hope every reader can benefit from it and work together to build a more complete prevention and control network against *Salmonella* to ensure global health.

Chenxi Huang
School of Life Science,
Hubei University,
Wuhan, China

Section 1

Disease Transmission and Management from One Health Perspective

Chapter 1

Salmonella Disease with a Health Management Perspective

Emine Kübra Dindar Demiray and Burak Sayar

Abstract

Salmonella infections represent a significant global health challenge, primarily due to their transmission through contaminated food and water, which affects individuals across all demographics. *Salmonella* spp., which are Gram-negative, rod-shaped pathogens, are responsible for various illnesses, ranging from gastroenteritis to more severe conditions such as typhoid fever. The prevalence of *Salmonella* infections exhibits global variability, significantly influenced by factors such as sanitation standards, food handling practices, and the robustness of public health infrastructure. Diagnosis typically involves culture analysis of stool, blood, or tissue samples, with treatment options complicated by increasing antibiotic resistance. Prevention and control measures emphasize food safety, public health education, and stringent hygiene practices. The chapter further elaborates on the significance of a multidisciplinary approach in health management to combat *Salmonella* infections effectively, including enhancing food safety inspections, expanding education programs, and improving laboratory capacities for infection control. The challenges of managing *Salmonella* are compounded by the pathogen's ability to cause outbreaks, the growing issue of antimicrobial resistance, and the necessity for effective health policies and regulations to reduce infection risks.

Keywords: antibiotic resistance, food safety, healthcare management, global health, *Salmonella*

1. Introduction

Infections resulting from the ingestion of contaminated food and water present a significant health issue, impacting individuals across all age groups. The primary contributory factors to this public health challenge are the consumption of foods and water supplemented with pathogenic microorganisms, including bacteria, viruses, and parasites [1]. In the field of health management, the management of infectious diseases, especially those caused by foodborne pathogens, is a critical issue that requires careful attention. In this context, the prevention and detection of *Salmonella* infections constitute a key public issue worldwide. This chapter provides a comprehensive overview of the etiology, epidemiology, and strategies of *Salmonella* in the field of health management [2]. The prevalence and severity of these infections vary depending on geographical location, socioeconomic conditions, and public health infrastructure. In developing countries, inadequate sanitation and limited supplies of clean drinking water lead to a higher prevalence of *Salmonella* infections, while in developed countries, deficiencies in food handling and preparation are a major risk factor [3]. In this context, prevention, detection, and control of these infections are of great importance for health management professionals.

Salmonella infections are usually diagnosed by culture analysis of stool, blood, or infected tissue samples. Laboratory tests are vital for identifying the causative agent and selecting suitable treatment methods without resorting to inhalation [4]. Nowadays, increasing antibiotic resistance is another factor complicating the treatment of infections. Therefore, preventing antimicrobial resistance is a priority for health management. The prevention and control of *Salmonella* infections covers a broad spectrum, starting from individual hygiene practices to the implementation of food safety standards and public health education. By implementing these multilayered approaches, health management professionals play an important role in reducing infection rates. This process requires effective collaboration between the food industry, public health authorities, and consumers. Furthermore, the development and implementation of health policies and regulations are critical for reducing the risk of infection [2, 3].

From a health management perspective, taking a proactive approach in the fight against *Salmonella* is vital in preventing the spread of the disease and protecting public health. This includes not only treating existing infections but also developing strategies to prevent future outbreaks [5]. These strategies include strengthening food safety inspections, expanding public health education programs and building enhanced laboratory capacities for infection control. This study aimed to address in depth the complex nature of *Salmonella* infections and the challenges faced in managing these infections. It is intended to provide an in-depth resource for health management professionals, policy makers, specialists in infectious diseases and microbiology, public health workers, and all other interested individuals seeking information in this area.

2. *Salmonella* disease

Salmonella infections hold significant implications for global public health, particularly within underdeveloped nations [6]. It can cause epidemics. The rise of this disease due to inadequate hygiene standards and limited regulatory controls further increases concerns [7]. The most important factors affecting prevalence factors include providing clean drinking water, safe wastewater removal, and access to clean food and climatic conditions [8, 9]. Transmission occurs via the fecal-oral route. There are numerous animal species associated with *Salmonella* infections. Among these, especially poultry, sheep, goats, cattle, horses, pigs, reptiles, wild birds, rodents, and domestic pets are notable. Additionally, infected humans can serve as sources of transmission, especially contaminated with feces consumption of water and foods (egg yolk, mayonnaise, milk, cream, vegetables and fruits, ice cream, etc.) for illness. High levels of *Salmonella* spp. ingested orally overcome the acidity of the stomach and reach the small intestine, where pathogenic *Salmonella* adhere to the intestinal epithelium and continue to live and multiply and spread. The infection first manifests as intestinal inflammation (gastroenteritis) [10, 11]. It lasts from 2 to 7 days, and its severity can vary. In some cases, the bacteria causing diarrhea can enter the bloodstream, leading to sepsis, which necessitates prompt and effective antimicrobial treatment. However, sepsis rarely occurs [2]. For disease to manifest in healthy individuals, a microbial load ranging between 10^5 and 10^{10} organisms is typically necessary. This threshold may be lower in young children, elderly individuals, and individuals with immunodeficiency. Factors such as the number of bacteria, virulence attributes, host age, and host immune conditions play pivotal roles in disease pathogenesis [6, 12].

Salmonella spp., are Gram-negative, rod-shaped bacterial pathogens belonging to the Enterobacteriaceae family. This bacterium is motile, unencapsulated, generally aerobic and facultatively anaerobic, and can reproduce [6]. When there is only one pathogen (blood, cerebrospinal fluid, etc.), it can grow on blood agar, chocolate agar, but when mixed flora is present (feces, etc.), it can only be produced on selective media such as SS agar and bismuth sulfate agar. *Salmonella* spp., which are divided into two taxonomic families as *Salmonella enterica* and *Salmonella bongori,* have different serogroups according to their somatic (O) antigen. Different serotypes have been defined based on virulence (Vi) antigen and flagellar (H) antigen [13]. More than 2600 *Salmonella* serovars have been identified. These serovars are categorized into typhoidal and non-typhoidal types. Typhoidal serovars exhibit high adaptation to the human host and are exclusively transmissible through human-to-human contact, causing potentially life-threatening syndromes such as typhoid or paratyphoid fever. Most cases in Europe are considered imported and typically involve individuals returning from endemic countries. Non-typhoidal serovars, on the other hand, are recognized as zoonotic agents capable of transmission from animals and foods to humans. They are commonly found in the environment and can contaminate water and food sources [14].

2.1 *Salmonella* epidemiology

Salmonella is one of the top three bacteria that causes food-associated enteritis [15]. In addition, enteric fever, acute enteritis, bacteremia, and invasive infections can lead to different infection patterns, such as asymptomatic carriage [6]. In 2010, the Foodborne Disease Burden Epidemiology Reference Group (FERG) of the World Health Organization (WHO) documented that *Salmonella* resulted in approximately 180 million cases of illness and contributed to 298,496 fatalities [16–19].

Globally, non-typhoidal *Salmonella* (NTS) is responsible for approximately 94 million cases of gastroenteritis annually, resulting in approximately 155,000 deaths. Data from the SalmSurv surveillance network, supported by the World Health Organization for foodborne illnesses, indicate that *S. enteritidis and S. typhimurium* are implicated in about 80% of these human infections [20, 21]. Currently, within the European Union (EU), *Salmonella enteritidis, S. typhimurium,* and its monophasic variant are recognized as the primary serovars accountable for human disease. *Salmonella enteritis* was most common in 2019 with 32,010 cases (61.6%); in 2020 with 21,203 cases (63.1%); and in 2021 with 23,634 cases (64.6%) [14]. Both pathogens affect both humans and animals [22]. While many serovars are non-pathogenic to animals, they pose a significant pathogenic risk to humans.

From January to October of 2023, a total of 335 instances connected to this epidemic were documented across 14 countries within the European Union (EU) and the European Economic Area (EEA), in addition to the United Kingdom (UK) and the United States (US). Between the start of January and late October 2023, across 14 EU/EEA countries, the United Kingdom, and the United States, there have been 335 confirmed cases of *Salmonella enteritidis* ST11, categorized into three microbiological groups, affecting individuals of various ages. A significant portion of the cases reported eating chicken products, notably chicken kebabs, prior to falling ill. There were nine hospitalizations in three nations and one fatality in Austria, emphasizing the ability of the outbreak to cause serious, sometimes lethal, illness [23].

Salmonella disease has recently garnered attention due to its diverse modes of transmission, notably through the consumption of pork or dry dog food. Additionally, there has been a recent focus on Charcuterie meats as potential sources of *Salmonella* contamination [24]. In terms of foodborne transmission, outbreaks within the EU have provided compelling evidence that various foods serve as carriers. For example, eggs and egg products were implicated in 39 outbreaks, mixed foods in 24, bakery products in 15, pig meat and related items in 14, and vegetables, juices, and their derivatives in 11 outbreaks. Additionally, raw milk and dairy products made with raw milk, seafood products, and processed foods such as sweets and chocolate have also been identified as sources of foodborne outbreaks in the EU [14]. Understanding the various transmission routes of *Salmonella* is essential for effective epidemiological management. Such insights underscore the importance of comprehensive surveillance and control measures to mitigate the risk of *Salmonella* outbreaks associated with diverse food sources.

2.2 *Salmonella* disease risk factors

Salmonella infections can originate from diverse food sources including meats like chicken and turkey, dairy products, and even some processed foods. Animal-derived raw products tend to have a higher risk of contamination, particularly uncooked meats, eggs, milk, and shellfish. Ensuring food safety involves preventing contamination throughout the production process and implementing control measures at critical points to secure access to safe food for consumption [1]. While various animal-derived foods are recognized as potential sources of *Salmonella* spp. infections, pork specifically has gained attention as a high-risk vector. A risk assessment model focusing on the pork supply chain indicates the incidence rate of salmonellosis from pork consumption in China, highlighting distinct rates for males and females [25]. The model's sensitivity analysis pinpoints critical factors impacting the risk associated with pork, such as display temperature, display duration, and the concentration of *Salmonella* spp. present in pork at retail points [25]. Apart from animal products, there is another way of contamination that should not be ignored. Multidrug-resistant and highly virulent strains of *Salmonella* are found in the small-scale production and distribution networks of leafy green vegetables. Multidrug-resistant strains of *Salmonella* spp. were detected in 92.4% of the samples collected from the water-plant-food nexus [7]. Recent multistate outbreaks of *Salmonella* have been associated with a variety of food items that are not traditionally linked to the pathogen. Investigations have connected cases of illness to products such as flour, peanut butter, salami sticks, onions, prepackaged salads, peaches, and ground turkey. These incidents underscore the importance of vigilance across a broad spectrum of foodstuffs to prevent contamination and subsequent *Salmonella* infections [24].

People, notably young children under the age of five, elderly adults over 65, and those with compromised immune systems due to certain health conditions—like diabetes, liver or kidney diseases, and cancer—or as a result of their medical treatments, are at an elevated risk for experiencing severe infections caused by *Salmonella*. It is crucial for these vulnerable groups to practice heightened food safety precautions [24]. Immune system suppression poses a risk for salmonellosis. Cancer, systemic lupus erythematosus, sickle cell anemia, and human immunodeficiency virus (HIV) infection have been reported as risk factors for invasive salmonellosis. *Salmonella* spp. stands out as a significant pathogen causing foodborne illnesses that pose a risk to human health. However, the transmission of this bacterium is not limited to food; it can also spread through contaminated water, environmental contact, interpersonal interactions, and animals, including pets and those found in petting zoos, farms, fairs, and educational settings.

The occurrence of *Salmonella* infection typically increases in the summer months of June, July, and August, in contrast to winter, which shows seasonal differences [26].

2.3 *Salmonella* disease signs and symptoms

S. enterica has different clinical characteristics according to its defined serotypes. The most important of these clinical pictures are food-related enteritis, bacteremia and localized organ infections, enteric fever (typhoid fever) and asymptomatic carriage [13]. While salmonellosis generally resolves on its own within a week, fatalities have been documented, particularly among susceptible demographic groups. These groups include very young individuals, elderly individuals, and individuals with compromised immune systems, highlighting the need for heightened vigilance and care in managing the disease within these populations [27]. The length and intensity of the illness can vary depending on the specific bacterial strains involved, the quantity of bacteria present, and the immune response of the host. This variation underscores the importance of considering both pathogen and host factors in understanding and managing the disease [28]. The disease presents with different clinical features [6].

2.3.1 Acute enteritis

Salmonellosis is frequently identified by symptoms resembling stomach flu or gastroenteritis. Gastroenteritis and food poisoning due to *Salmonella* sometimes occur as sporadic cases, and sometimes as outbreaks in places where public meals are held, such as hospitals, schools, dormitory canteens, and restaurants [10].

Symptoms of this condition include nausea, vomiting, abdominal pain, and diarrhea that may be bloody, along with headaches and fever. Prolonged symptoms can lead to dehydration, a concern, particularly for vulnerable groups such as infants and elderly individuals. This emphasizes the need for prompt and appropriate hydration and medical intervention in affected individuals. It develops on average 6–72 hours (average 24 hours) after consumption of food contaminated with *S. enteritidis*, *S. choleraesuis*, and *S. typhimurium*. It may also be accompanied by muscle pain and weakness. It may cause mesenteric lymphadenitis, which can cause severe abdominal pain that can be confused with acute appendicitis [8, 9, 29].

2.3.2 Bacteremia and localized organ infections

S. typhi, S. typhimurium, S. choleraesuis, S. enteritidis, S. paratyphi A, B, and *C,* and *S. Heidelberg* have the ability to cause bacteremia by passing into the bloodstream. Although the frequency of transient bacteremia resulting from *Salmonella* gastroenteritis is not clearly known, it is thought to be approximately 1–5%. This rate increases in the neonatal period and in conditions such as HIV infection and sickle cell anemia [30, 31].

In some cases, signs of sepsis (fever, chills, sweating, myalgia, loss of appetite, and weight loss), as well as findings associated with focal suppurative infection such as meningitis and osteomyelitis, may be detected [8, 13].

2.3.3 Enteric fever (typhoid fever)

S. typhi is primarily associated with enteric fever, a condition that may also be caused by *S. paratyphi A, B,* and *C,* though less frequently. The illness is marked by symptoms such as fever, loss of appetite, headaches, fatigue, muscle pain, and constipation, in addition to other nonspecific symptoms. In severe cases where the infection leads to septicemia or meningitis, it poses a fatal risk, underscoring the severity of these bacterial infections and the importance of timely diagnosis and treatment [18]. The process of diagnosis solely through clinical manifestations presents difficulties, owing to the symptomatic resemblance with various infections that are widespread in endemic regions. These include viral (such as dengue and influenza), parasitic (such as malaria, typhus, and leishmaniasis), and bacterial (such as brucellosis and tuberculosis) diseases. This overlap in symptomatology underscores the necessity for comprehensive diagnostic approaches to accurately identify the underlying cause of the patient's symptoms [31]. Microorganisms are taken via the fecal-oral route approximately 10–14 days after fever, fatigue, loss of appetite, dry cough, and abdominal pain begin. The incubation duration may vary between 3 and 30 days depending on the inoculum amount, bacterial virulence, age, and host characteristics, such as the immune response and vaccination history. The clinical manifestations of *S. paratyphi A*, *B*, and *C* have a milder clinical picture. It causes paratyphoid fever [13].

On physical examination, relative bradycardia can be detected depending on the severity of fever. Maculopapular rashes (rose spot) may occur in 7–10 days of the disease, lasting for 2–3 days, usually by palpation of the abdomen. The spleen is softly and sensitively palpated in the first 2 weeks of the disease. Patients with a toxic appearance (dull facial expression, rusty tongue, and foul mouth odor) are detected [29, 31]. Normochromic normocytic anemia, leukopenia, and neutropenia may be detected. Thrombocytopenia, coagulation abnormalities, and hypofibrinogenemia can be observed in cases of DIC development [6].

According to US data, the frequency of non-typhoid *Salmonella* (NTS) infections has increased sixfold in the last 40 years. The reason for this is the rapid growth of the ready-made food industry rapid growth, disruption of flora as a result of improper use of antibiotics, and the increase of immune deficiency in the population. Due to increasing HIV infection, malnutrition, and poor hygiene conditions, non-typhoidal *Salmonella* (NTS) have emerged as a leading cause of mortality and morbidity among children in Africa [9].

2.3.4 Asymptomatic carriage

Enteric fever or *Salmonella* gastroenteritis can result in the continuation of *Salmonella* shedding through feces for a duration of up to 5 weeks. Individuals who persist in shedding *Salmonella* after 1 year are considered asymptomatic chronic carriers. The likelihood of becoming a carrier is notably higher in individuals with gallbladder stones, those under the age of five, women, elderly individuals, and patients with gastrointestinal tract cancers. Moreover, in patients suffering from tuberculosis of the urinary system or *Schistosoma* infection, the excretion of *Salmonella* in urine may also be observed [8]. The existence of Mary Mallon, a healthy carrier, has made this disease more recognizable [32].

2.4 Diagnosis of *Salmonella* disease

The diagnosis of *Salmonella* infection is accomplished through a careful assessment of patients' clinical symptoms and the application of microbiological tests. This process typically involves culture analysis of samples of blood, feces, or other body fluids that may be infected. A rapid and accurate diagnosis is critical for determining appropriate treatment approaches and reducing potential risks to public health.

2.4.1 Diagnosis of Salmonella enteritis

A common cause of food-associated bacterial diarrhea is *Salmonella* infection. During stool examination, leukocytes and erythrocytes are detected. In rare cases, it is necessary to perform rectoscopy for differential diagnosis. In these patients, macroscopic mucosal edema, hyperemia, mucosal fragility, and bleeding foci can be observed. Definitive diagnosis is made by producing the agent. *S. enteritidis* and *S. typhimurium* are the most frequently produced agents. In blood and suppurative infections, aspirate liquid culture and Gram staining must be done [8, 13]. For the causes of invasive diarrhea (such as *Shigella*, *Yersinia enterocolitica, Clostridium difficile,* and enterohemorrhagic *E. coli*) and acute appendicitis, intestinal perforation should be considered in the differential diagnosis [8, 9].

2.4.2 Diagnosis of enteric fever

For the definitive diagnosis of enteric fever, laboratory testing is essential. Despite the recognized significance of enteric fever for over a century, the search for a single "ideal" laboratory diagnostic biomarker has yet to yield a universally accepted solution. Diagnostic tests should be performed when the fever is ≥38°C and lasts ≥3 days [4, 33].

Definitive diagnosis is made by testing the causative agent in blood, bone marrow, feces, or duodenal liquid culture. For diagnosing enteric fever, in addition to the use of blood and bone marrow for bacterial culture, additional biological materials such as rose spots, duodenal bile, stool, and urine might be cultured to detect *Salmonella*. Culturing from rose spots, although noninvasive and offering approximately 60% sensitivity, is less frequently applicable as these spots appear in only 1–30% of enteric fever cases, indicating the necessity for a broad approach in sample collection for effective disease identification [34]. The detection of nucleic acids employs polymerase chain reaction (PCR) techniques to amplify serovar-specific DNA of *Salmonella* for diagnostic purposes. This method's significant advantage lies in its swift processing time. PCR is particularly beneficial as it is capable of identifying DNA from *Salmonella*, whether the bacteria are alive, dead, or both, thereby offering a versatile and efficient diagnostic tool [33]. Serological detection of *S. enterica* serovars utilizes the Kauffman-White classification, although its specificity for diagnosing enteric fever is not perfect due to antigen commonality among different serovars. Despite this, serological tests are simple and rapid, making them crucial for prompt disease management in resource-limited endemic regions. The Gruber-Widal test is a prevalent serological method in such settings, detecting agglutination of bacterial antigens by specific antisera. For enhanced accuracy, it is advised to conduct the test twice, once during the acute phase and again during convalescence, seeking a fourfold increase in antibody titers for a positive result [31, 35, 36]. However, it should not be forgotten that chronic liver disease, infection with other Gram-negative agents or hypogammaglobulinemia, can result in false negatives; in disseminated tuberculosis and malaria, false positive can be detected [13]. The diagnosis of bacteremia and localized organ infections is similar to that of enteric fever. In addition, the radiological appearance specific to the involved organ and bacteriological growth in that organ is also detected [13].

2.4.3 Diagnosis of asymptomatic carriage

Long-term asymptomatic carriage, exceeding 1 year, is rare. Considering that transient excretion following infection is frequent, whereas prolonged asymptomatic carriage is infrequent, and the absence of substantiated evidence endorsing antibiotic treatment for chronic carriers, conducting follow-up stool cultures for the impacted children and their relatives is deemed unnecessary [6]. In some cases, however, it may be recommended to collect three consecutive stool samples at least one month apart after completion of antibiotic therapy [11]. Although outbreaks of typhoid fever that occur in underdeveloped countries are usually linked to contaminated water, the most common mode of serotype *typhi* transmission in the United States is the asymptomatic carrier [37]. Diagnosis is usually made with stool samples examined consecutively. Chronic carrier screening commonly utilizes the Vi antigen [33, 38].

2.5 *Salmonella* disease treatment

Treatment of *Salmonella* infection often varies depending on the severity of the disease and the general health status of the affected individual. Mild cases may often require no specific treatment, and symptomatic treatment with adequate hydration may often be sufficient. However, severe or high-risk patients may require antibiotic therapy and supportive care in hospital.

2.5.1 Salmonella enteritis *treatment*

Dehydration and electrolyte irregularity are recommended for rapid evaluation. For treatment, usually oral or IV rehydration and correction of electrolytes is sufficient. Antibiotic treatment in asymptomatic chronic illness is not recommended because it increases the carrier rate. However, for three-month-old babies, elderly individuals, and people in high-risk groups (patients with immunodeficiency, HIV, chemotherapy, who use immunosuppressive drugs and have sickle cell anemia patients) should be used empirical treatment due to the risk of bacteremia. Third-generation cephalosporins and ciprofloxacin may be preferred for empirical treatment [13, 39]. Multiple multidrug-resistant (MDR) *Salmonella* spp. are being isolated at increasing rates. Therefore, according to culture and antibiotic sensitivity testing, appropriate antibiotic treatment should be continued [31, 40].

2.5.2 Treatment of bacteremia and localized organ infections

The third generation cephalosporin treatment for bacteremia thought to be due to NTS should be preferred. Ciprofloxacin, ceftriaxone, and cefotaxime are first option, and cefixime and azithromycin are used as alternatives. Chloramphenicol, ampicillin, and trimethoprim/sulfamethoxazole are used only if susceptibility is detected. Cephalosporin, quinolone, and azithromycin combination is available in severe cases [13]. If there is resistance to all first and second choice drugs, carbapenem and tigecycline can be used [40, 41].

The prescribed duration for antibiotic therapy varies depending on the condition being treated: 14 days for bacteremia, 4–6 weeks for acute osteomyelitis, and a minimum of 4 weeks for meningitis. While some research suggests a 2-month treatment period for uncomplicated osteomyelitis, a duration of at least 3 months is advised for chronic or complex infections. The decision to discontinue antibiotic therapy should be based on the observed therapeutic response. Additionally, it is imperative to conduct clinical studies to ascertain the optimal duration of treatment for these conditions [8, 42, 43].

2.5.3 Treatment of enteric fever

Early diagnosis and treatment of typhoid is important. Patients should be given rest, adequate fluid, and electrolyte support. Paracetamol or ibuprofen may be given as antipyretic treatment. Nutrition should be continued, unless intestinal complications develop the patient with soft, light, easily digestible foods. Usually within 3days after starting antibiotic treatment, the fever begins to drop, and the patient starts to feel better. Adequate treatment takes an average of 2–3 weeks. Although treatment is needed, 1–4% of patients relapse within the second week [6].

2.5.4 Treatment of asymptomatic carriage

High-dose ampicillin for 6 weeks or treatment with a fluoroquinolone for 4 weeks is implemented. Cholecystectomy is recommended for people with gallstones whose decolonization cannot be achieved with this treatment. Treatment with ampicillin is recommended for 10 days before the operation and 4 weeks after the operation. Chronic carriers in the food industry are not recommended for public health purposes [8, 13].

2.6 New diagnosis, treatment approaches, and vaccines

The scientific community is actively engaged in the search for new acute biomarkers specifically tailored to enteric fever, with the goal of distinguishing these cases from other infectious diseases. The quest for ideal biomarkers for enteric fever, which are yet to be validated clearly, continues. The absence of an animal model that fully mirrors the infection lifecycle of typhoidal *Salmonella* strains poses a significant obstacle in understanding the disease pathogenesis and in the search for new biomarkers. There is a critical need for biomarkers that are detectable early in the infection, can reveal resistance to treatment, and accurately identify acute cases from subclinical infections or chronic carriers, especially in regions where the disease is endemic [33].

The ELISA technique and a method known as the TPTest, which evaluates IgA levels against the membrane components of *S. typhi* and *S. paratyphi* using ALS samples, have demonstrated a specificity range of 78–97% and a sensitivity of 100% in identifying the bacteria. This indicates a high level of accuracy in detecting these pathogens, demonstrating the potential of these methods for diagnosing infections caused by these *Salmonella* serovars [44, 45]. The TPTest offers a significant advantage by being able to distinguish between the acute phase of infection and the convalescent phase. This capability renders it a critical diagnostic tool within endemic regions for accurately diagnosing the disease status of individuals [46]. Further enhancements to these biomarkers could enable their advancement into point-of-care (POC) compatible rapid diagnostic tools, offering a more immediate and accessible option for disease detection.

While traditional culture-based methods for diagnosing enteric fever show high specificity, they are limited by their sensitivity and extended processing times. PCR techniques offer quicker results but require specialized skills and equipment. Serological tests, such as the Widal test, are rapid and suitable for point-of-care use in endemic regions, though their specificity and sensitivity are moderate. Efforts to identify distinctive bacterial and host biomarkers for enteric fever, aiming to differentiate between the stages of infection, have led to high-throughput studies. However, no single biomarker has yet been identified as ideal due to insufficient data [33]. Advanced technologies like proteomics, transcriptomics, and metabolomics, despite their potential, advanced technologies such as proteomics, transcriptomics, and metabolomics are often not feasible in endemic areas due to infrastructure limitations. Consequently, there is an urgent requirement for the development of more cost-effective, simpler methods or the adaptation of these advanced technologies into more accessible forms. The primary challenge in diagnosing *S. typhi* and *S. paratyphi* infections in such regions lies in the need for diagnostic solutions that are both affordable and user-friendly. Furthermore, future research should also focus on devising diagnostic methods capable of identifying healthy or asymptomatic carriers of these pathogens, despite the complexities involved in this endeavor [33].

The initial development of typhoid vaccines dates back to the late 19th century. There are various vaccines developed for this purpose, but none of them provide long-term protection, and also not 100% effective and protective antibody levels can be achieved with annual booster doses every 3–5 years. Two vaccines licensed in the United States are accessible for *Salmonella typhi*: an oral attenuated live virus vaccine and an injectable polysaccharide vaccine. The Vi-polysaccharide typhoid vaccine is administered to adults over 2 years of age via a single intramuscular dose. The effect continues for 3 years, and its effectiveness is 70–80%. The Ty21a vaccine is an oral live typhoid vaccine. It is recommended that three doses be administered 2 days apart. Antibiotics should not be taken, 7 days before and after vaccination. It is not recommended for children under 6 years old or pregnant and for those with primary immunodeficiency or who are receiving immunosuppressive medication. This vaccine provides protection for 5 years, and its effectiveness is 67–82%. The conjugated forms of the Vi typhoid vaccine and tetanus vaccine are produced in India. This vaccine (Typbar-TCV) can also be used at 9 months of age [43, 44]. The FDA does not recommend the typhoid vaccine for routine immunization of individuals in the United States [47].

2.7 Protection

Salmonella infections are an important public health issue that should not be neglected. A significant portion of patients have this infection severe enough to require hospital inpatient treatment. *Salmonella* species cause epidemics and epidemics may cause economic losses. This situation reveals the importance of minimizing losses due to the disease, preventing the disease, and taking these prevention measures [10, 48].

All *Salmonella* infections can be prevented only by breakage in the fecal-oral chain. For this purpose, hand washing is the first step. Local administrators are responsible for providing infrastructure services and provide clean domestic water and food to people and animals. Another important issue is preventing contamination with animal feces during cutting and storage of all products, effective cooking, and pasteurization. In short, clean food consumption is required [10, 13, 48]. Food contaminated with *Salmonella* or other harmful germs usually looks, tastes, and smells normally. Therefore, it is important to know how to prevent infection. The use of pets by the veterinarian regularly makes the pets healthy. When a pet is healthy, household is also protected [24]. Risky people, such as employees in laboratory, who has household exposure to a chronic *Salmonella typhi* carrier, living in endemic areas and who travel to these regions for more than 2 weeks, can also be protected from typhoid with the *S. typhi* vaccine [49, 50].

3. Health management and food safety

The global impact of foodborne diseases is significant, affecting nearly one in ten individuals annually and resulting in the loss of 33 million healthy life years. These diseases disproportionately affect young children, with diarrheal diseases being the most prevalent outcome of consuming unsafe food. Annually, 550 million people are afflicted, including 220 million children under the age of five. Among the leading causes of diarrheal diseases worldwide, *Salmonella* ranks as one of the top four pathogens [51].

In this context, the magnitude of the burden of foodborne diseases has become an important focus of governments and health policy makers. Therefore, ensuring food safety is not only an ongoing problem from a historical perspective but also a critical public health priority today.

Food safety has been an important issue throughout history in terms of its impact on human health. Since the beginning of recorded history, unsafe food consumption has been a threat to human health. Many of the food security challenges we face today are not new but have historical roots. On a global scale, governments are developing and implementing various strategies to maximize the security of the food supply. However, in both developed and developing countries, the emergence of foodborne diseases remains a major public health challenge. Five key points have been identified for safer food. These are as follows [52]:

- It should be kept clean
- Raw and cooked foods should be separated
- It should be cooked well
- Store food at safe temperatures
- Use safe water and raw materials

4. Conclusions

Salmonella infection poses a significant public health challenge, as this bacterial disease is one of the most common causes of food poisoning and can affect millions of people worldwide. *Salmonella* is a major foodborne pathogen and can cause gastroenteritis, typhoid fever, and paratyphoid fever, particularly through contaminated water and food. The severity of *Salmonella* infections can range from mild cases of gastroenteritis to life-threatening systemic diseases, increasing the impact of the infection on public health [6, 9, 15]. Prevention and control of disease due to this bacterium is possible through effective food safety practices, public health surveillance, and public awareness efforts. Early diagnosis and treatment of *Salmonella* infections are vital to protect public health and reduce the burden of foodborne illness. Therefore, reducing the burden of *Salmonella* infections on public health is possible not only through individual and societal measures but also through proactive and holistic approaches of health systems. In this context, controlling the infection and preventing its spread should be at the center of health management strategies, and coordination of multidisciplinary efforts in this field should be a priority. *Salmonella* is a complex and multilayered problem in terms of health management.

The etiology, epidemiology, and health system impact of *Salmonella* infections emphasize the need for a comprehensive and multidisciplinary approach to the management of this disease [53]. Prevention, early diagnosis, and effective treatment of infections are vital to protect public health and utilize health resources effectively. In the fight against *Salmonella*, strict enforcement of food safety and hygiene standards plays a fundamental role in preventing the spread of the disease. This requires continuous awareness raising and supervision at the food production and processing stages, in consumers' food preparation and storage practices, and in public health education programs. Furthermore, the development and implementation of health policies and regulations are critical in the prevention and control of these infections [2, 9, 53].

One of the biggest challenges in the diagnosis and treatment of *Salmonella* infections is antibiotic resistance. This is a growing problem worldwide and requires the development of new strategies for the management and treatment of infections [54]. Health management professionals need to play a leading role in the fight against antimicrobial resistance and promote research and innovation in this field. Furthermore, rapid and accurate diagnosis of infections increases the effectiveness of infection control programs and reduces the burden on health systems. From a health management perspective, another important factor in combating *Salmonella* infections is public health education. Informing the public about how infections are transmitted, prevention and control are critical steps in preventing the spread of these diseases. Health management professionals should develop effective communication strategies to disseminate this information and raise public awareness.

In conclusion, the management of *Salmonella* disease requires constant attention and innovation in the field of health management. It is important to recognize the complexity of *Salmonella* infections and the importance of health management of these infections. Health management professionals, policy makers, and public health workers should continuously strive for effective prevention, diagnosis, and treatment of this disease and keep abreast of evolving knowledge in this field. This approach will contribute to the protection of both individual and public health and help prevent future *Salmonella* outbreaks. Within the scope of this study, certain recommendations were developed. These recommendations are as follows:

- It is important to strengthen international cooperation to enhance global monitoring of *Salmonella* and data sharing. This can contribute to a better understanding of disease patterns and prevent disease spread.

- Organizing community-wide awareness-raising and education programs on food safety can increase individuals' awareness of safe food consumption.

- International food safety standards should be effectively implemented in food production, processing, distribution, and consumption stages.

- Increasing the use of existing vaccines against *Salmonella* and investing in research and development on vaccine development, diagnostic methods, and treatment approaches will be an important step in the fight against the disease.

- Adopting multidisciplinary approaches to solving food safety problems can produce effective solutions by combining the knowledge and experience of different disciplines.

- Continuous review and improvement of hygiene and sanitation standards in food businesses can reduce the risk of transmission of pathogens such as *Salmonella*.

- Establishing improved monitoring and reporting systems for the early detection of foodborne diseases is vital to prevent the spread of disease.

- Improving labeling and information practices for consumers can help people make more informed decisions about their food choices.

- Harmonizing safety standards in the international food trade can limit the international spread of foodborne diseases.

- Increasing the frequency and scope of food safety inspections is an important factor in reducing risks in food businesses.

- Encouraging innovations in food processing technologies can enable safer processing and storage of food.

- Strictly controlling the use of antibiotics in the production of animal products can help prevent antibiotic resistance and thus control resistant pathogens such as *Salmonella*.

Conflict of interest

The authors declare no conflict of interest.

Thanks

We would like to thank all the authors and the editorial team involved in the preparation of this resource.

Author details

Emine Kübra Dindar Demiray[1]* and Burak Sayar[2]

1 Afyonkarahisar Public Hospital Infectious Diseases and Clinical Microbiology, Afyonkarahisar, Turkey

2 Bitlis Eren University Vocational School of Social Sciences, Bitlis, Turkey

*Address all correspondence to: e.kubradindar@hotmail.com

References

[1] Şireli U, Cengiz G. Gıda Kaynaklı Mikrobiyal Hastalıkların 2019 Verilerine Genel Bakış: Geleneksel Derleme. Turkiye Klinikleri Journal of Veterinary Sciences. 2023;**14**(1):28

[2] Galán-Relaño Á, Valero Díaz A, Huerta Lorenzo B, Gómez-Gascón L, Mena Rodríguez Mª Á, Carrasco Jiménez E, et al. Salmonella and salmonellosis: An update on public health implications and control strategies. Animals. 2023;**13**(23):3666

[3] Ngogo FA, Joachim A, Abade AM, Rumisha SF, Mizinduko MM, Majigo MV. Factors associated with Salmonella infection in patients with gastrointestinal complaints seeking health care at regional hospital in Southern Highland of Tanzania. BMC Infectious Diseases. 2020;**20**:1-8

[4] Andrews JR, Ryan ET. Diagnostics for invasive salmonella infections: Current challenges and future directions. Vaccine. 2015;**33**:C8-C15

[5] Mkangara M. Prevention and control of human Salmonella enterica infections: An implication in food safety. International Journal of Food Science. 2023:1

[6] Yücel E. *Salmonella* enfeksiyonları, tanı ve tedavisi. Klinik Tıp Pediatri Dergisi. 2020;**12**(3):133-139

[7] Kgoale DM, Duvenage S, Du Plessis EM, Gokul JK, Korsten L. Serotype distribution, antimicrobial resistance, virulence genes, and genetic diversity of Salmonella spp. isolated from small-scale leafy green vegetable supply chains in South Africa. Journal of Food Protection. 2024;**87**(1):100195

[8] Bhutta ZA. *Salmonella*. In: Kliegman RM, Behrman RE, Jenson HB, Stanton BF, editors. Nelson Textbook of Pediatrics. 21st ed. Philadelphia: Saunders Elsevier Inc; 2020. pp. 2235-2248

[9] Gordon M. Invasive non-typhoidal *Salmonella* disease-epidemiology, pathogenesis and diagnosis. Current Opinion in Infectious Diseases. 2011;**24**:484-489

[10] Cesare AD. *Salmonella* in foods: A reemerging problem. Advances in Food and Nutrition Research. 2018;**86**:137-179

[11] Khanam F, Darton TC, Meiring JE, Kumer SP, Kumar BP, Bhuiyan MAI et al. Salmonella Typhi stool shedding by patients with enteric fever and asymptomatic chronic carriers in an endemic urban setting. The Journal of Infectious Diseases. 2021;**224**(Supplement_7):S759-S763

[12] Raffetelu M, Wilson P, Winter SE, Baumler AJ. Clinical pathogenesis of typhoid fever. Journal of Infection in Developing Countries. 2008;**2**(4):260-266

[13] Ochoa TJ, Cleary TG. *Salmonella*. In: Feigin RD, editor. Textbook of Pediatric Infectious Diseases. 8th ed. Philadelphia: Saunders Elsevier Inc; 2019. pp. 1066-1081

[14] European Food Safety Authority and European Centre for Disease Prevention and Control (EFSA and ECDC). 2022. The European Union One Health 2021 Zoonoses Report. EFSA Journal 2022;**20**(12):7666, 273 pp. DOI: 10.2903/j.efsa.2022.7666

[15] Majowicz SE, Musto J, Scallan E, et al. The global burden of nontyphoidal *Salmonella* gastroenteritis. Clinical Infectious Diseases. 2010;**50**:882-889

[16] Wen SC, Best E, Nourse C. Non-typhoidal *Salmonella* infections in children: Review of literature and recommendations for management. Journal of Paediatrics and Child Health. 2017;**53**(10):936-941

[17] Scallan E, Hoekstra RM, Mahon BE, Jones TF, Griffin PM. An assessment of the human health impact of seven leading foodborne pathogens in the United States using disability adjusted life years. Epidemiology and Infection. 2015;**143**(13):2795-2804. DOI: 10.1017/S0950268814003185

[18] World Health Organization (WHO). Food safety [Internet]. Geneva: WHO; 2020. Available from: https://www.who.int/news-room/fact-sheets/detail/food-safety [Accessed: Apr 4, 2022]

[19] Ehuwa O, Jaiswal AK, Jaiswal S. *Salmonella*, food safety and food handling practices. Food. 2021;**10**(5):907

[20] Stanaway JD, Parisi A, Sarkar K, Blacker BF, Reiner RC, Hay SI, et al. The global burden of non-typhoidal *Salmonella* invasive disease: A systematic analysis for the global burden of disease study 2017. The Lancet Infectious Diseases. 2019;**19**:1312-1324

[21] World Health Organization. WHO Estimates of the Global Burden of Foodborne Diseases: Foodborne Disease Burden Epidemiology Reference Group 2007-2015. Geneva, Switzerland: World Health Organization; 2015. Available from: WHO website [Accessed: January 11, 2024]

[22] Kirk MD, Pires SM, Black RE, Caipo M, Crump JA, Devleesschauwer B, et al. World Health Organization estimates of the global and regional disease burden of 22 foodborne bacterial, protozoal, and viral diseases, 2010: A data synthesis. PLoS Medicine. 2015;**12**:e1001921

[23] European Centre for Disease Prevention and Control, European Food Safety Authority, 2023. Three clusters of Salmonella Enteritidis ST11 infections linked to chicken meat and chicken meat products – 26 October 2023. DOI: 10.2903/sp.efsa.2023.EN-8388

[24] Centers for Disease Control and Prevention (CDC). Food Safety [Internet]. Atlanta: CDC; 2024. Available from: https://www.cdc.gov/salmonella [Accessed: Jan 17, 2024]

[25] Zhong J, Zhou G, Yang Y, Sun X, Zhang H, Qu X, et al. Quantitative risk assessments of Salmonella spp. in domestic pork in China. Brazilian Journal of Microbiology. 2024;**55**:681-688

[26] Brown M, Eykyn SJ. Non-typhoidal *Salmonella* bacteraemia without gastroenteritis: A marker of underlying immunosuppression. Review of cases at St. Thomas' hospital 1970-1999. The Journal of Infection. 2000;**41**:256-259

[27] Turgeon P, Ng V, Murray R, Nesbitt A. Forecasting the incidence of *Salmonellosis* in seniors in Canada: A trend analysis and the potential impact of the demographic shift. PLoS One. 2018;**13**:e0208124

[28] Waddington CS, Darton TC, Jones C, Haworth K, Peters A, John T, et al. An outpatient, ambulant-design, controlled human infection model using escalating doses of *Salmonella Typhi* challenge delivered in sodium bicarbonate solution. Clinical Infectious Diseases. 2014;**58**:1230-1240

[29] Bula-Budas FJ, Rathore MH, Maraqa NF. *Salmonella* infections in childhood. Advances in Pediatrics. 2015;**62**:29-58

[30] Shkalim V, Amir A, Samra Z, Amir J. Characteristics of nontyphi *Salmonella* gastroenteritis associated with bacteremia in infants and young children. Infection. 2012;**40**:285-289

[31] Kumar P, Kumar BS. Enteric fever. Indian Journal of Pediatrics. 2017;**84**(3):227-230

[32] Entis P. Food Safety: Old Habits, New Perspectives. Washington DC: ASM Press; 2007

[33] Neupane DP, Dulal HP, Song J. Enteric fever diagnosis: Current challenges and future directions. Pathogens. 2021;**10**(4):410

[34] Gilman RH, Terminel M, Levine MM, Hernandez-Mendoza P, Hornick RB. Relative efficacy of blood, urine, rectal swab, bone-marrow, and rose-spot cultures for recovery of *Salmonella typhi* in typhoid fever. Lancet. 1975;**1**:1211-1213

[35] Levine MM, Grados O, Gilman RH, Woodward WE, Solis-Plaza R, Waldman W. Diagnostic value of the Widal test in areas endemic for typhoid fever. The American Journal of Tropical Medicine and Hygiene. 1978;**27**:795-800

[36] House D, Chinh NT, Diep TS, Parry CM, Wain J, Dougan G, et al. Use of paired serum samples for serodiagnosis of typhoid fever. Journal of Clinical Microbiology. 2005;**43**:4889-4890

[37] Crump JA, Karlsson MS, Gordon MA, Parry CM. Epidemiology, clinical presentation, laboratory diagnosis, antimicrobial resistance and antimicrobial treatment of invasive salmonella infections. Clinical Microbiology Reviews. 2015;**28**(4):901-937

[38] Lanata CF, Levine MM, Ristori C, Black RE, Jimenez L, Salcedo M, et al. Vi serology in detection of chronic *Salmonella typhi* carriers in an endemic area. Lancet. 1983;**2**:441-443

[39] Sirinavin S, Garner P. Antibiotics for treating salmonella gut infections. The Cochrane Database of Systematic Reviews. 2000;**1**(2):CD001167. DOI: 10.1002/14651858.cd001167. PMID: 10796610

[40] Crump JA. Typhoid fever and the challenge of nonmalaria febrile illness in sub-saharan Africa. Clinical Infectious Diseases. 2012;**54**:1107-1109

[41] Robert-Jan H, Goessens WHF, Wilfrid VP, et al. *Salmonella* subtypes with increased MICs for azithromycin in travelers returned to The Netherlands. Emerging Infectious Diseases. 2014;**20**:705-708

[42] Santos EM, Sapico FL. Vertebral osteomyelitis due to *Salmonellae*: Report of two cases and review. Clinical Infectious Diseases. 1998;**27**:287-295

[43] Huang ZD, Wang CX, Shi TB, Wu BJ, Chen Y, Li WB, et al. *Salmonella* osteomyelitis in adults: A systematic review. Orthopaedic Surgery. 2021;**13**(4):1135-1140

[44] Islam K, Sayeed MA, Hossen E, Khanam F, Charles RC, Andrews J, et al. Comparison of the performance of the TPTest, Tubex, Typhidot and Widal immunodiagnostic assays and blood cultures in detecting patients with typhoid fever in Bangladesh, including using a Bayesian latent class modeling approach. PLoS Neglected Tropical Diseases. 2016;**10**:e0004558

[45] Sheikh A, Bhuiyan MS, Khanam F, Chowdhury F, Saha A, Ahmed D, et al. *Salmonella enterica serovar Typhi*-specific immunoglobulin a antibody responses in plasma and antibody in lymphocyte supernatant specimens in Bangladeshi

patients with suspected typhoid fever. Clinical and Vaccine Immunology. 2009;**16**:1587-1594

[46] Khanam F, Sheikh A, Sayeed MA, Bhuiyan MS, Choudhury FK, Salma U, et al. Evaluation of a typhoid/paratyphoid diagnostic assay (TPTest) detecting anti-*Salmonella* IgA in secretions of peripheral blood lymphocytes in patients in Dhaka, Bangladesh. PLoS Neglected Tropical Diseases. 2013;**7**:e2316

[47] Britto CD, Wong VK, Dougan G, Pollard AJ. A systematic review of antimicrobial resistance in *Salmonella enterica serovar Typhi*, the etiological agent of typhoid. PLoS Neglected Tropical Diseases. 2018;**12**(10):e0006779

[48] Mac Fadden DR, Bogoch II, Andrews RJ. Advances in diagnosis, treatment, and prevention of invasive *Salmonella* infections. Current Opinion in Infectious Diseases. 2016;**29**:453-458

[49] Van Camp RO, Shorman M. Typhoid vaccine. In: Stat Pearls. Treasure Island (FL): StatPearls Publishing; Jan 2024. Available from: https://www.ncbi.nlm.nih.gov/books/NBK470571/

[50] Milligan R, Paul M, Richardson M, Neuberger A. Vaccines for preventing typhoid fever. Cochrane Database of Systematic Reviews. 2018;(5):CD001261. DOI: 10.1002/14651858.CD001261.pub4 [Accessed: April 01, 2024]

[51] World Health Organization. *Salmonella* (non-typhoidal) [Internet]. Geneva: WHO; 2024. Available from: https://www.who.int/news-room/fact-sheets/detail/*salmonella*-(non-typhoidal) [Accessed: Feb 7, 2024]

[52] World Health Organization. Five Keys to Safer Food Manual [Internet]. Geneva: WHO; 2006. Available from: https://iris.who.int/bitstream/handle/10665/43546/9789241594639_eng.pdf?sequence=1

[53] Chousalkar KK, Sexton M, McWhorter A, Hewson K, Martin G, Shadbolt C, et al. Salmonella Typhimurium in the Australian egg industry: Multidisciplinary approach to addressing the public health challenge and future directions. Critical Reviews in Food Science and Nutrition. 2017;**57**(12):2706-2711

[54] Montoro-Dasi L, Lorenzo-Rebenaque L, Marco-Fuertes A, Vega S, Marin C. Holistic strategies to control Salmonella infantis: An emerging challenge in the European broiler sector. Microorganisms. 2023;**11**(7):1765

Chapter 2

The Role of Birds in Salmonellosis

Amina Tufail and Yusra Ashfaq

Abstract

The main research question explores the role of birds in *Salmonella* transmission dynamics, highlighting the zoonotic potential for disease spread from birds to humans. Investigating bird salmonellosis uncovers a complex interaction of factors, including birds' feeding habits and ecological behaviors, that facilitate disease dissemination. Utilizing epidemiological evidence, genomic analysis, and case studies, the research elucidates how salmonellosis is transmitted among birds and to humans. Significant findings identify garden birds as major *Salmonella* reservoirs, reveal genetic similarities between avian and human *Salmonella* strains, and underscore the role of environmental contamination. Notably, species like the Eurasian siskin and greenfinch are significant carriers, with bird-feeding practices and feeder congregations accelerating transmission. The research emphasizes a comprehensive management approach to salmonellosis, involving public education, wildlife management, and strategic interventions to reduce risks. It calls for improved hygiene practices in human-bird interaction zones and targeted bird population surveillance for *Salmonella*. Understanding birds' role in salmonellosis epidemiology is vital for developing effective public health strategies, enhancing food safety, and wildlife conservation, and underscoring the interconnectedness of human, animal, and environmental health in tackling zoonotic diseases.

Keywords: *Salmonella* transmission, avian vectors, public health, environmental contamination, hygiene practices

1. Introduction

1.1 Definition of salmonellosis

Bird salmonellosis is a complicated and multifaceted problem that affects equally domestic and wild bird populations. Primary transmission occurs through contaminated food, but environmental and direct contact vectors also play a role in the spread of the illness. The bacteria *Salmonella enterica* are the primary drivers of this disease, which can take numerous forms and frequently cause severe gastrointestinal illness, fever, diarrhea, stomach cramps, and, tragically, high rates of death. Serovars such as *Salmonella Typhimurium* are mainly retained in both domestic and wild birds [1]. Notably, certain serovars, such as *Salmonella Typhimurium*, have adapted to persist in bird populations, with phage types such as DT40 and DT56v identified as prevalent in these hosts [1]. The zoonotic potential of *Salmonella* makes its extensive presence in bird populations a major public health problem and an ongoing concern for wildlife conservation efforts. Zoonosis, or bird flu, can spread from birds to humans through many channels, such as coming into contact directly with sick birds, touching bird baths or feeders, or having a pet that comes into contact with sick animals [2].

Salmonellosis significantly threatens wild birds, particularly passerine species such as finches and sparrows, as evidenced by their elevated mortality rates [3]. This disease's spread is notably facilitated by behaviors common among these birds, such as congregating in large numbers at feeding sites. These gatherings, often around bird feeders provided by humans, serve as prime venues for *Salmonella* transmission, effectively turning them into hotspots for the disease. The resulting rapid spread not only jeopardizes bird conservation efforts by affecting the health and survival of these species but also increases the zoonotic risk, bridging the gap between wildlife diseases and human infections. This risk is especially pronounced in areas where there is close contact between humans and wild birds [4].

The connection between *Salmonella* infections in wild bird populations and human cases has been firmly established through extensive epidemiological research, underscoring the zoonotic nature of this pathogen. In light of this, raising awareness about salmonellosis and promoting good hygiene practices in places frequented by wild birds are crucial steps toward mitigating the risk of transmission but also increasing zoonotic risk [5]. Furthermore, proactive measures to monitor and control the spread of *Salmonella* among bird populations are vital, particularly given the emerging status of this disease. Over the last four decades, there has been a notable increase in the incidence of salmonellosis among wild birds, a trend likely propelled by the increase in artificial feeding practices. The birds most at risk are those that feed on the ground or in contaminated water sources, including those that consume both plant and animal matter, highlighting the need for vigilant management of the environments we share with these animals [2].

1.2 Key insights into salmonellosis spread and prevention

The study conducted by Lawson et al. [4] provides critical insights into the role of garden birds as primary reservoirs in the transmission dynamics of salmonellosis. This research, focused on England and Wales, revealed the significant involvement of garden birds in the spread of zoonotic bacteria responsible for human salmonellosis cases. Lawson et al. [4] reported that garden birds are significant reservoirs for bacteria that cause salmonellosis in humans. This study highlighted the connection between the presence of these birds and the incidence of salmonellosis cases in human populations. Research on the species-specific aspects of transmission revealed that certain garden bird species including Greenfinches (*Carduelis chloris*) and House Sparrows (*Passer domesticus*) are more likely to harbor and spread the *Salmonella* bacteria. This finding is crucial for understanding the epidemiology of salmonellosis and designing targeted control measures. The study also considered the environmental and ecological factors that contribute to the prevalence of salmonellosis in garden birds. Factors such as bird-feeding habits, migration patterns, and interactions with human environments were explored to understand how they influence the spread of disease. The identification of garden birds as primary reservoirs for salmonellosis-causing bacteria has significant public health implications. This suggests the need for better management practices in urban and suburban gardens and parks where human interaction with these birds is frequent. Lawson et al. proposed recommendations for controlling the spread of salmonellosis from garden birds to humans. These include public awareness campaigns, safe bird-feeding practices, and enhanced surveillance of bird populations for *Salmonella* infections. This research also revealed the broader impact of salmonellosis on wildlife and domestic animal populations, suggesting that garden birds could be a source of infection for a wide range of animals, thereby impacting biodiversity and animal health.

The study by Mather et al. [6] on the genomic analysis of *Salmonella enterica* in wild passerines provides profound insights into the transmission dynamics of salmonellosis. The research identified genetic similarities between the *Salmonella* strains found in wild birds and those that infect humans. These similarities suggest that the same or closely related strains of *Salmonella* are capable of infecting both birds and humans, pointing to a shared pathway or common source of infection. The researcher conducted a genomic analysis of the *Salmonella* strains isolated from birds. They found notable genetic overlaps with strains that infect humans. This finding implies that the same or closely related strains of *Salmonella* are capable of infecting multiple species. This shared genetic profile suggests a common evolutionary pathway or a shared environmental reservoir that facilitates cross-species transmission or similar mechanisms to cause disease in different hosts. This can include similarities in how bacteria invade host cells, evade the immune system, or produce toxins. Another aspect of the genetic basis could be related to antimicrobial resistance. The presence of similar resistance genes in both bird and human *Salmonella* strains is a matter of concern for treatment and public health, as it suggests the potential for the spread of resistance traits across species. The findings of the genetic study conducted by Mather et al. [6] contribute to molecular epidemiology, which helps trace the origins and spread of infections by examining the genetic makeup of pathogens. Understanding the genetic similarities allows for better tracking of infection sources and pathways. The genetic analysis tools used in the study are crucial for public health surveillance. These methods enable health authorities to identify and respond to outbreaks more effectively by understanding.

The 2014 study by Krawiec et al. [7] significantly advanced our understanding of how certain garden birds, specifically the Eurasian siskin and greenfinch, act as significant carriers and reservoirs for *Salmonella* spp., playing a critical role in the transmission dynamics of salmonellosis. The transmission cycle involves the birds acquiring the *Salmonella* bacteria, likely from their environment or through interaction with other infected birds or animals. Once infected, these birds can shed the bacteria through their feces, which can then contaminate their surroundings, including food sources, water, and areas where they congregate. This transmission cycle is particularly concerning because it provides multiple opportunities for the spread of *Salmonella* to other birds, wildlife, and potentially to humans, especially in settings where there is close interaction between humans and these bird species. This research has considerable implications for public health and wildlife management, especially in areas where Eurasian siskins and greenfinches are common, as it focuses on controlling the spread of *Salmonella* within bird populations. This study enables more targeted surveillance and intervention efforts. Such efforts can include monitoring bird populations for signs of infection, implementing measures to mitigate the spread of bacteria in bird habitats, and educating the public about safe practices for bird feeding and handling. Furthermore, the study contributes significantly to the One Health approach, emphasizing the interconnected nature of human, animal, and environmental health.

2. Epidemiology of Salmonellosis

2.1 Prevalence and incidence

Salmonellosis is a worldwide health issue that impacts millions of people each year. Its occurrence varies by region and is affected by factors such as food hygiene practices and interactions between animals and humans. This variety is a result of the many situations under which *Salmonella* can propagate and impact populations [8]. Enhanced food handling and sanitation measures have reduced salmonellosis in certain areas, but in other regions, interactions between humans and animals contribute to disease transmission. Comprehending these dynamics is essential for creating efficient public health interventions to lessen the impact of salmonellosis on a global scale.

2.1.1 Subclinical salmonellosis in wild birds near pig farms

Andrés et al. [9] reported that *Salmonella* was more prevalent in birds near pig farms. The study detected *Salmonella* in 1.85% of bird fecal samples, with *Salmonella Typhimurium* being the most prevalent serotype. This indicates that pig farms may amplify *Salmonella* infection among wild birds, with non-migratory birds being particularly affected [9].

2.1.2 Salmonellosis in garden birds and human transmission in England and Wales

Lawson et al. [4] reported that garden birds are a significant source of human salmonellosis in England and Wales. This study linked the phage types of *Salmonella* found in birds to those found in humans, supporting the hypothesis that garden birds act as the primary reservoir of these zoonotic bacteria Lawson et al. [4].

2.1.3 Salmonella *in wild birds and retail food in Singapore*

Aung et al. [10] found a 0.99% prevalence of *Salmonella* in wild birds in Singapore, suggesting that avian *Salmonella* isolates are subject to less antimicrobial selection pressure than those from food sources. This study helps to elucidate the epidemiology of human salmonellosis in wild birds [10].

2.1.4 Salmonella typhimurium *O:4-12 infection in Norway linked to wild birds*

Kapperud et al. [11] reported that the epidemic strain of *Salmonella typhimurium* O:4-12 in Norway often causes fatal salmonellosis among wild passerine birds. The study indicated a clear seasonal pattern and an epidemiologic link between avian and human cases [11].

Salmonellosis is a global health concern, with millions affected annually. Its incidence varies geographically and is influenced by factors such as food hygiene practices and animal-human interactions [12].

2.2 Common *Salmonella* serovars

The study conducted by Şik et al. [13] in Turkey provided a comprehensive analysis of *Salmonella* serovars across various animal species. They examined 1047

Salmonella spp. strains collected between 2015 and 2020, revealing a diverse range of serovars: 19 serogroups and 75 different serovars. The most commonly isolated serovar was *Salmonella Infantis*, which is primarily found in chickens. Other notable findings included *Salmonella montevideo* in calves, *Salmonella darle* in tortoises, and *Salmonella typhimurium* in lambs. Of particular interest was the identification of *Salmonella* Hessarek in wild birds, emphasizing the prevalence of this serovar in avian species. This study highpoints the intricate distribution of *Salmonella* serovars in various animal populations, highlighting the need for targeted surveillance and control strategies [13].

In Central Spain, Martín-Maldonado et al. [14] investigated the presence and antimicrobial resistance of *Salmonella* in urban birds. They found that 12.3% of the 300 sampled birds were *Salmonella* positive, with white storks feeding in landfills showing the highest prevalence. This study identified zoonotic serovars such as *Salmonella enteritidis* and *S. typhimurium*, including their monophasic variants. A significant finding was the high resistance rate to antibiotics, with 40.5% of *Salmonella* strains being resistant to ciprofloxacin, nalidixic acid, and colistin. These results demonstrate the potential role of urban birds as reservoirs and disseminators of antimicrobial-resistant *Salmonella* strains, posing a public health concern [14].

Fu et al. [15] conducted a study in the United States to examine the resistance of *Salmonella enterica* strains in wild birds. They sequenced 375 *S. enterica* strains collected from wild birds in 41 states from 1978 to 2019. The dominant serovar found was *Typhimurium*, comprising 68.3% of bird isolates. Notably, the proportions of isolates identified as multi-antimicrobial resistant (1.87%) or multi-heavy metal resistant (1.87%) were low. These findings suggest that wild birds in the United States do not serve as significant reservoirs for multi-resistant *S. enterica* strains, although continuous surveillance is necessary due to the potential transmission of resistant bacteria to humans and domestic animals [15].

Wales and Lawes [16] focused on *Salmonella* Gallinarum, a serovar that specifically affects avian species like such as chickens and turkeys. This serovar causes invasive and septicemic diseases, often leading to high mortality rates. *S. Gallinarum* has two biovars, *Pullorum* and *Gallinarum*, both of which cause severe disease in poultry but with different transmission dynamics. This study underscores the importance of clean breeding practices for controlling this pathogen, though sporadic outbreaks still occur even in countries with stringent control measures [16].

Smith et al. [17] identified the bar-tailed godwit, a migratory shorebird, as a reservoir for *Salmonella* Hvittingfoss in Australia. This serovar was responsible for a foodborne outbreak in 2016–2017 linked to tainted cantaloupes. The study found a genetically similar strain of S. Hvittingfoss in the godwit population, highlighting the potential role of migratory birds in the long-distance transmission of foodborne pathogens. This discovery is significant in understanding the dynamics of *Salmonella* transmission and the potential environmental reservoirs of foodborne pathogens [17].

Numerous serovars of *Salmonella* exist, with some such as *Typhimurium* and *Enteritidis* being particularly prevalent in human infections [18].

2.2.1 Human health impact

The disease can range from mild gastroenteritis to severe systemic infections, especially in vulnerable populations such as children and immunocompromised individuals (**Figure 1**) [19].

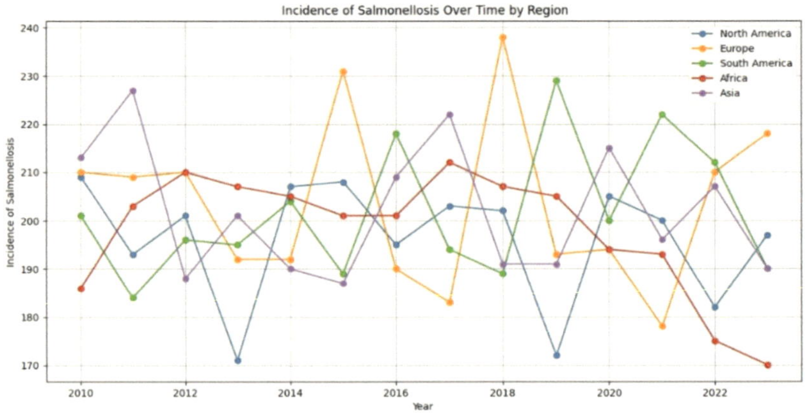

Figure 1.
Graphs showing the incidence of salmonellosis over time in different region.

3. Zoonotic transmission

Zoonotic diseases, also known as zoonoses, are infections that can be transmitted from animals to humans, either through direct contact with infected animals, indirectly through environmental exposure, or by consuming contaminated food products. These diseases encompass a wide range of health implications, from mild symptoms to severe illnesses, and are a significant concern in both public health and veterinary medicine.

One notable zoonotic disease is salmonellosis, which is caused by *Salmonella* bacteria. *Salmonella* is widely distributed in the animal kingdom, particularly in birds and reptiles, as well as in mammals and amphibians. Research by Ellis in 1969 highlighted birds as significant reservoirs for *Salmonella*, emphasizing the risk of transmission to humans. *Salmonella* infections can lead to gastrointestinal illnesses and more severe conditions, especially in individuals with weakened immune systems [12].

The primary sources of human infection with *Salmonella* include contaminated food products such as poultry, eggs, meat, and dairy products derived from infected animals. Direct contact with infected animals or their fecal matter is another significant source of transmission, posing a particular risk in settings like farms, veterinary clinics, or even in households with pets like turtles and birds. Additionally, environmental contamination is an increasingly recognized vector for salmonellosis. Contaminated water sources, soil, or surfaces can harbor *Salmonella* bacteria, broadening the scope of potential exposure beyond animal products and direct animal contact [20].

In 2011, Hoelzer, Moreno Switt, and Wiedmann noted out the growing importance of environmental contamination, including contaminated produce, as a source of human salmonellosis. This acknowledgment necessitates a comprehensive approach to controlling outbreaks, which includes monitoring potential animal reservoirs and environmental contamination, in addition to food safety measures [21].

The study of zoonotic transmission, particularly of *Salmonella*, is crucial for developing preventive strategies and managing public health risks associated with these diseases. Understanding the diverse transmission pathways helps in implementing targeted interventions to reduce the incidence and impact of zoonotic diseases (**Figure 2**).

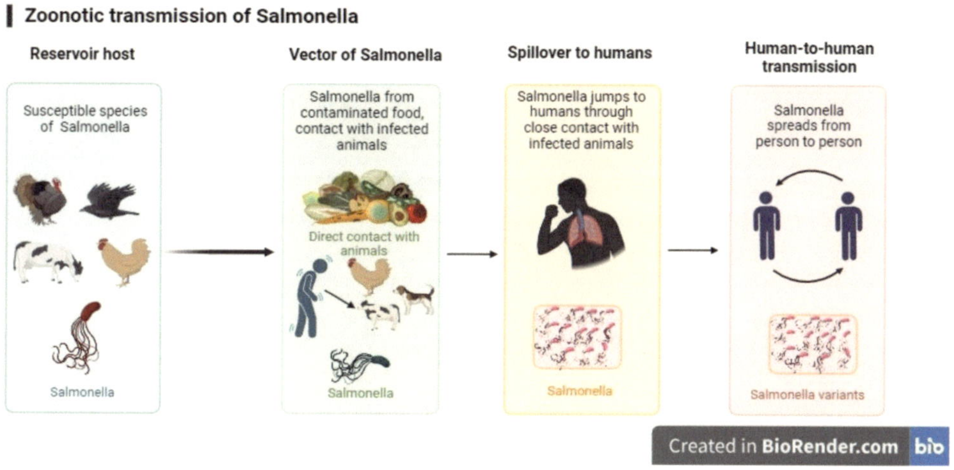

Figure 2.
Flowchart depicting the zoonotic transmission cycle of salmonellosis.

4. Birds as carriers of *Salmonella*

Birds, encompassing a wide range of wild and domesticated species, play a significant role as reservoirs for the *Salmonella* bacteria. These avian hosts can harbor and disseminate various serovars of this bacterium, contributing to the spread of salmonellosis.

4.1 Avian reservoirs of *Salmonella*

The presence of *Salmonella* in birds is a well-documented phenomenon. Studies have shown that both wild and domesticated bird populations can carry and transmit different serovars of *Salmonella*. Lawson et al. [4] specifically highlighted the role of garden birds as sources of human salmonellosis in England and Wales, indicating the direct link between avian carriers and human infections. This relationship underscores the importance of birds in the epidemiology of *Salmonella*, not only as carriers but also as potential sources of transmission to other species, including humans [4].

4.2 Factors contributing to *Salmonella* carriage in birds

Several factors influence the likelihood of birds carrying *Salmonella*. These include:

1. Diet: the type of food consumed by birds can affect their risk of *Salmonella* infection. Birds that feed on a diet rich in seeds or those scavenging for food in human-inhabited areas or farms may be more likely to encounter and ingest *Salmonella* bacteria [7].

2. Environmental exposure: birds living in or near contaminated environments, such as agricultural lands with livestock or areas with poor sanitation, have higher chances of coming into contact with *Salmonella* [22].

3. Interactions with other animals or humans: birds that interact with other infected animals or with humans, particularly in settings such as poultry farms,

Species	Commonly identified serovars	References
Eastern Bluebird	*Typhimurium*, other serovars	[22]
Chipping Sparrow	*Typhimurium*, other serovars	[22]
Cattle Egret (*Bubulcus ibis*)	*Typhimurium*, other serovars	[22]
House Sparrow (*Passer domesticus*)	*Typhimurium*, other serovars	[22]
Fish Crows (Corvus ossifragus)	*Typhimurium*, Panama, other serovars	[22]
Cormorants, Pelicans, Herons	*Typhimurium*, other serovars	[7]
Gulls and Terns	*Typhimurium*, other serovars	[7]
Eurasian Siskins	*Typhimurium*	[7]
Greenfinches	*Typhimurium*	[7]
Asian Pied Starling (*Gracupica contrasta*)	Not specified	No relevant data provided
House Crow (*Corvus splendens*)	Not specified	No relevant data provided
Seabirds	Panama, *Typhimurium*, other serovars	[7]

Table 1.
List of bird species and their associated Salmonella *serovars.*

petting zoos, or households, can acquire and spread *Salmonella*. The interaction does not need to be direct or indirect contact through contaminated surfaces or equipment, which can also facilitate transmission [22].

Meade and Barrow [23] discussed these factors, emphasizing the complexity of *Salmonella* transmission in avian populations. They pointed out that the interaction of these factors creates a dynamic environment where the risk of *Salmonella* carriage can vary significantly among different bird populations and environments [23].

In summary, the role of birds as carriers of *Salmonella* is a multifaceted issue influenced by various ecological and environmental factors. Understanding these dynamics is crucial for developing strategies to control the spread of salmonellosis and protect public health (**Table 1**).

5. *Salmonella* in wild birds

Salmonella in wild birds is an area of significant interest due to the role these birds play in the transmission of the bacterium. Wild birds, including various garden and marine species, have been identified as carriers of *Salmonella*, and they can act as vectors, facilitating the spread of this pathogen [24].

5.1 Prevalence in wild bird populations

The prevalence of *Salmonella* in wild bird populations varies depending on numerous factors, including geographical location, bird species, and local environmental conditions. A study by Tardone et al. [25] highlighted the presence of *Salmonella* in both garden and marine bird populations, demonstrating the widespread nature of this issue. These birds can pick up *Salmonella* from various sources and potentially spread it over long distances, especially in the case of migratory specie [25].

5.2 Risk factors for wild birds

Several factors contribute to the risk of wild birds harboring and spreading *Salmonella*:

1. Foraging habits: birds that forage in areas where they can come into contact with *Salmonella*, such as agricultural lands or places with poor waste management, are at a higher risk of infection.

2. Habitat overlap with humans and domestic animals: birds living near human habitations or domestic animals are more likely to encounter sources of *Salmonella*. This is particularly true for birds that frequently live in urban areas or farm environments.

3. Environmental contamination: areas contaminated with *Salmonella*, whether through industrial activities, sewage, or other means, can serve as hotspots for birds to acquire the bacterium.

Minette [26] discussed these risk factors, indicating that the likelihood of wild birds carrying *Salmonella* is influenced by their interactions with both natural and anthropogenic environments. For example, birds that feed in contaminated areas or interact with domestic animals are more likely to become carriers of *Salmonella* [26].

In summary, wild birds play a significant role in the epidemiology of *Salmonella* due to their prevalence in diverse bird populations and the various risk factors that facilitate bacterial transmission. Understanding these dynamics is crucial for assessing the potential public health risks and for developing effective strategies to mitigate the spread of *Salmonella* from wild birds to humans and other animals (**Figure 3**).

Figure 3.
Map showing the prevalence of salmonellosis in wild birds across different continents [27].

6. *Salmonella* in domesticated birds

6.1 *Salmonella* in poultry

One of the most significant reservoirs for *Salmonella* is poultry flocks, and the products of poultry flocks are frequently the source of human infections. *Salmonella* can be passed on to people through contact with infected birds, feces, or contaminated goods such as eggs and meat. *Salmonella* can also be spread through cross-contamination [22].

6.2 Measures to prevent the spread of *Salmonella* in poultry farms

Salmonella poses a notable threat to chicken farms, and it is imperative to implement control measures to prevent its dissemination. Several strategies can be implemented to manage *Salmonella* in poultry farms:

6.2.1 Measures to ensure biosecurity

Implementing biosecurity measures is crucial to prevent the entry of *Salmonella* into the farm. Implementing biosecurity measures is crucial for preventing *Salmonella* from entering farms. These methods include regulating entry to farms, sterilizing equipment and trucks, and preventing interactions between wild birds and poultry [9]. Ensuring limited entry to farms is a crucial concern, and the provision of shower and changing room amenities for both workers and tourists can effectively mitigate the transmission of *Salmonella*.

6.2.2 Vaccination

Vaccination is a highly efficient and economical approach to mitigate the consequences of clinical *Salmonella* infections in chickens. Vaccines can decrease the occurrence of *Salmonella* in groups of chicken and hinder the transmission of the pathogen to humans [28, 29].

6.2.3 Flock surveillance for infection

Surveillance of chicken flocks is essential for identifying the presence of *Salmonella* infection. Sampling of regular chicken flocks should be monitored to oversee the production chain, and prompt measures should be implemented upon detection of *Salmonella* [28]. Identifying the source of the illness is crucial for preventing the dissemination of bacteria.

6.2.4 Sanitization and sterilization

Thoroughly cleaning and disinfecting poultry houses and equipment is crucial for preventing the transmission of *Salmonella*. It is essential to exercise caution when cleaning and disinfecting. Additionally, before replenishing the facility, it is necessary to conduct a bacteriological investigation [30].

6.3 *Salmonella*-free birds

The crucial initial step in preventing illnesses is introducing birds free from *Salmonella* [31]. Incoming avian specimens must exhibit optimal health conditions,

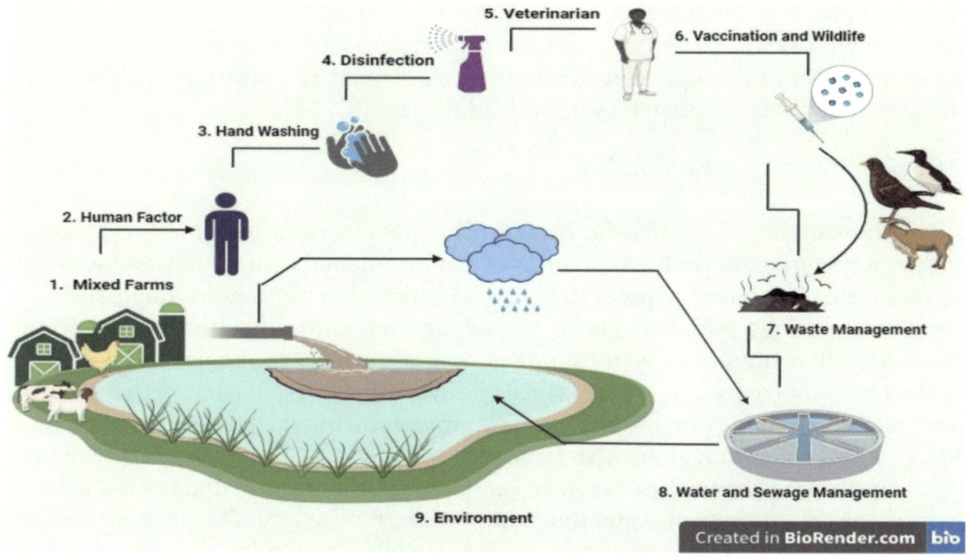

Figure 4.
Diagram of biosecurity measures in a poultry farm setup. 1. The biosecurity plan should include the separation of domestic species, their removal, and testing for any diseases. 2. As a biosecurity precaution, the human factor attempts to minimize indirect transmission by discouraging equipment sharing and supporting thorough disinfection of shared equipment. Additionally, visitors should be provided with personal protective equipment when in close contact with animals. 3 and 4. Utilizing detergents for on-farm washing will enhance the efficiency of debris and biofilm removal. 5. The veterinarians must maintain a sanitary atmosphere. 6. Preventive environmental measures on farms should encompass the implementation of physical barriers or fences, monitoring programs, vaccination protocols, and mandated testing of natural fauna. 7. Wastes from diseased animals and any biological wastes from wild animals must undergo proper disposal procedures. Water management and environmental contamination in the range of 8–9 restrict animals from accessing natural water sources such as streams and creeks, which could potentially contain infections [32].

and the relocation of poultry groups after their production cycle should exclusively be permitted for slaughter. Poultry farms can mitigate the occurrence of *Salmonella* in their flocks and hinder the transmission of the infection to humans by implementing these management methods (**Figure 4**).

7. Transmission routes

Salmonella, a prevalent and possibly lethal bacterium, can infect humans through different means, including direct interaction with birds, pollution of the environment, and transmission through food. Understanding these transmission pathways is essential for implementing efficient preventive strategies. New research has provided insight into the many mechanisms by which *Salmonella* can be transmitted from birds to humans.

7.1 Transmission of diseases from birds to humans

Salmonella can be transmitted to humans by direct contact with birds or their excrement. This transmission mode is most pertinent in environments where people are close to avian species, such as in domestic poultry-rearing or bird-feeding endeavors. Research has demonstrated that *Salmonella* can be transferred among hens, cattle, and pigs through infected feed or the environment in free-range situations. Moreover, wild avian species have the potential to transmit *Salmonella* to humans by direct contact or through exposure to their excrement and saliva, particularly in settings where birds congregate, such as bird feeders [33, 34].

7.2 Environmental contamination

The transmission of *Salmonella* from birds to people, resulting in indirect human contact, is a significant problem in terms of environmental contamination. Both wild and domestic birds have the potential to contaminate diverse surroundings, such as agricultural settings, with *Salmonella*, so endangering human health. By deposition, *Salmonella* can contaminate water supplies, soil, or surfaces near avian habitats. Studies have emphasized the significance of environmental pollution in the transmission of *Salmonella* from birds to humans, underscoring the need to reduce environmental exposure in regions with large bird populations. Moreover, the presence of *Salmonella* in water sources has raised concerns due to its potential to affect the microbiological integrity of water and compromise food safety. The primary sources of *Salmonella* contamination on produce include soil, manure, irrigation water, and exposure to reptiles, birds, or other small animals. It has garnered significant attention as a potential reservoir of *Salmonella* contamination, and research has linked it to *Salmonella* contamination [22, 35].

7.3 Transmission of pathogens through food

Contaminated poultry products are a significant means of transmitting *Salmonella* to humans. Poultry goods, including eggs and meat, can potentially be infected with *Salmonella*, which presents a substantial hazard to consumers. Ensuring the appropriate management, preparation, and preservation of chicken products is crucial for preventing the spread of *Salmonella* through food. Research has shown that *Salmonella* is commonly found in birds that live in urban areas, indicating there is a risk of spreading the bacteria through the consumption of contaminated chicken products [36, 37].

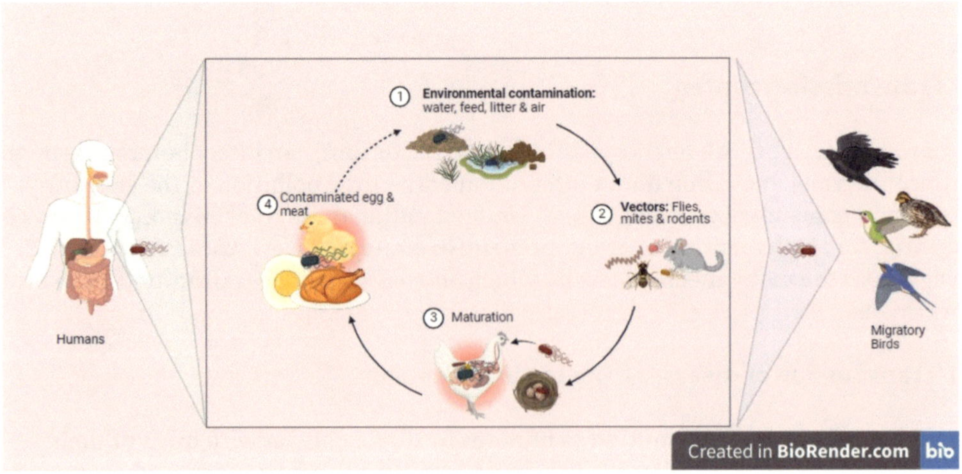

Figure 5.
Infographic showing different transmission routes from birds to humans.

To summarize, *Salmonella* can be transmitted from birds to humans through different pathways, such as direct contact, environmental pollution, and transmission through contaminated food. It is crucial to implement measures restricting exposure to various transmission modes, such as maintaining excellent hygiene, following safe food handling practices, and addressing environmental contamination. Various actions are necessary to decrease the risk of *Salmonella* infection in humans (**Figure 5**).

8. *Salmonella* strains and serovars in birds

8.1 Common Serovars in birds

Serovars like *Typhimurium* and *Enteritidis* are commonly found in birds [38]. *Salmonella enterica* serovars, including *Typhimurium* and *Enteritidis*, are frequently present in various bird species, whether they are wild or domestic. These serovars are especially problematic since they can cause illness in birds and humans. These serovars have been classified as host-specific and linked to salmonellosis in avian species, such as pigeons and passerines. Moreover, the rise of antibiotic-resistant *Salmonella* strains in avian species is becoming increasingly problematic, as it could compromise the efficacy of antimicrobial therapy for avian and human illnesses [7, 39].

Studies have demonstrated that various forms of *Salmonella*, such as *Typhimurium*, can infect a wide range of hosts. Nevertheless, they may also adapt to specific hosts, with different lineages linked to particular bird species. An investigation of 131 S. *Typhimurium* isolates obtained from wild birds in the United States revealed that isolates from various taxonomic host groups, such as passerine birds, water birds, and larids, belonged to three separate lineages. This suggests that genetic characteristics are associated with adaptability to different hosts [40].

8.2 Antimicrobial resistance in Avian-associated *Salmonella*

The emergence of antimicrobial-resistant strains in birds is a growing concern [41]. The increase in antimicrobial-resistant strains of *Salmonella* in avian species is a burgeoning concern that substantially impacts the well-being of animals and humans. Poultry-associated *Salmonella* isolates have developed antimicrobial resistance (AMR) due to the extensive use of antimicrobial agents on farms. Consequently, the emergence of antibiotic-resistant *Salmonella* strains has presented a formidable obstacle in effectively treating *Salmonella* infections in both avian and human populations [8].

An analysis of antibiotic resistance in *Salmonella* spp. obtained from poultry worldwide revealed a significant prevalence of resistance (60.7%) in *Salmonella* isolates against routinely employed antibiotics in poultry farming, including nalidixic acid and ampicillin. The study also identified significant antibiotic resistance levels toward streptomycin, amoxicillin/clavulanic acid, and trimethoprim/sulfamethoxazole. The results emphasize the extensive occurrence of antibiotic resistance in *Salmonella* strains obtained from poultry, emphasizing the immediate requirement for efficient approaches to control the dissemination of antimicrobial resistance in bird populations [42].

The appearance of *Salmonella* strains that are resistant to antimicrobial drugs in broiler chickens has also sparked worries about the consequences of using antimicrobial agents in the production of food animals. The detection of antibiotic-resistant *Salmonella* in meat products has increased worries regarding the potential ramifications of antimicrobial usage in the poultry sector. As a result, there has been a demand for the adoption of antimicrobial agents and the development of alternative approaches to decrease antibiotic dependence in chicken farming [43].

8.3 *Salmonella* strain in birds reported from different countries

Research has been conducted in several areas to examine the occurrence and genetic traits of *Salmonella* in birds that live in the wild. This has yielded vital knowledge on the specific strains and serovars linked to salmonellosis in birds. Notable discoveries derived from the given sources include the following:

8.4 Incidence and serotypes in Poland

A study conducted in Poland has revealed that *Salmonella* ser. *Typhimurium* as a causative agent of salmonellosis in pigeons and passerines. This finding confirms that certain wild bird species serve as reservoirs for *Salmonella* serotypes, particularly *Salmonella* ser. *Typhimurium* [7].

8.5 Molecular characterization in the United States

A study conducted in the United States revealed that *Salmonella enterica* serovar *Typhimurium* isolates obtained from wild birds exhibited unique lineages that were determined by the kind of bird. Some of these isolates displayed genetic characteristics indicating adaptability to their specific hosts. Genetic sequencing of 131 *S. Typhimurium* samples obtained from wild birds in 30 states across the United States was performed. The samples were collected between the years 1978 and 2019 [39].

8.6 Incidence and variety in environments of fresh produce

Studies have demonstrated a significant amount of diversity in bird feces, with the identification of up to three serovars of *Salmonella* from a single sample, indicating a wide range of variations. The culture-based investigation confirmed the presence of a wide range of serovars among the isolates, thereby indicating a high level of diversity [22].

8.7 *Salmonella* in wild birds in Bangladesh

A study conducted in Bangladesh found that 65% of *Corvus splendens* (house crow) and 67% of *Gracupica contra* (Asian Pied Starling) were infected with *Salmonella*. This highlights the potential public health importance of salmonellosis transmitted by birds (**Figure 6**) [45].

9. Public health concerns

The transmission of salmonellosis from birds can cause substantial health hazards, particularly during outbreaks. The increasing prevalence of antibiotic-resistant strains in avian populations is problematic since it can compromise the efficacy of antimicrobial therapy for avian and human illnesses [46]. Public health must track and manage outbreaks associated with avian sources, as they have the potential to cause the transmission of *Salmonella* to people and other animals [47].

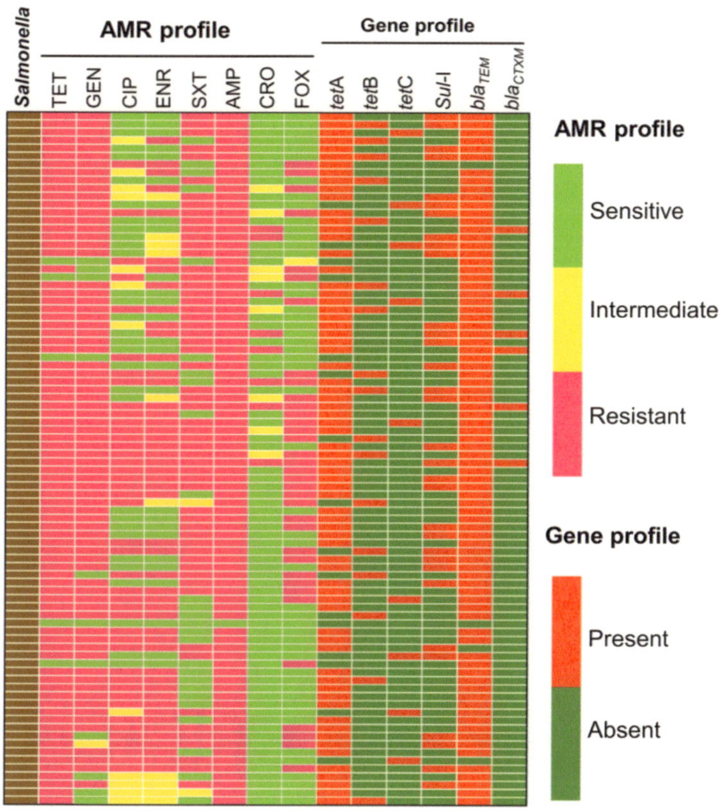

Figure 6.
Chart illustrating the antimicrobial resistance patterns of avian Salmonella *strains. The antimicrobial resistance phenotypes and genotypes of the* Salmonella *isolates were analyzed for the distribution of resistance to different antibiotics. The abbreviations used are as follows: TET for tetracycline, GEN for gentamicin, CIP for ciprofloxacin, ENR for enrofloxacin, SXT for sulfamethoxazole/trimethoprim, AMP for ampicillin, CRO for ceftriaxone, and FOX for cefoxitin [44].*

9.1 Outbreaks and surveillance

Tracking and managing outbreaks linked to avian sources are critical for public health [38]. CDC [48] *Salmonella* infections associated with avian sources have been documented in different environments, such as in backyard poultry farming and bird-feeding activities. In the United States, health authorities have examined numerous outbreaks of *Salmonella*, including *Enteritidis*, Indiana, Infantis, Mbandaka, and *Typhimurium*, which have been connected to contact with backyard poultry. These widespread occurrences emphasize the significance of closely observing and monitoring *Salmonella* infections to prevent their dissemination [48].

9.2 Impact on public health

Salmonellosis from avian sources poses significant health risks, especially in outbreak scenarios [49]. Avian-related salmonellosis can result in various symptoms, such as diarrhea, abdominal cramps, and fever, which can potentially lead to severe illness and mortality in some instances. The presence of harmful microorganisms in poultry meat and its by-products poses a significant threat to public health, as it can

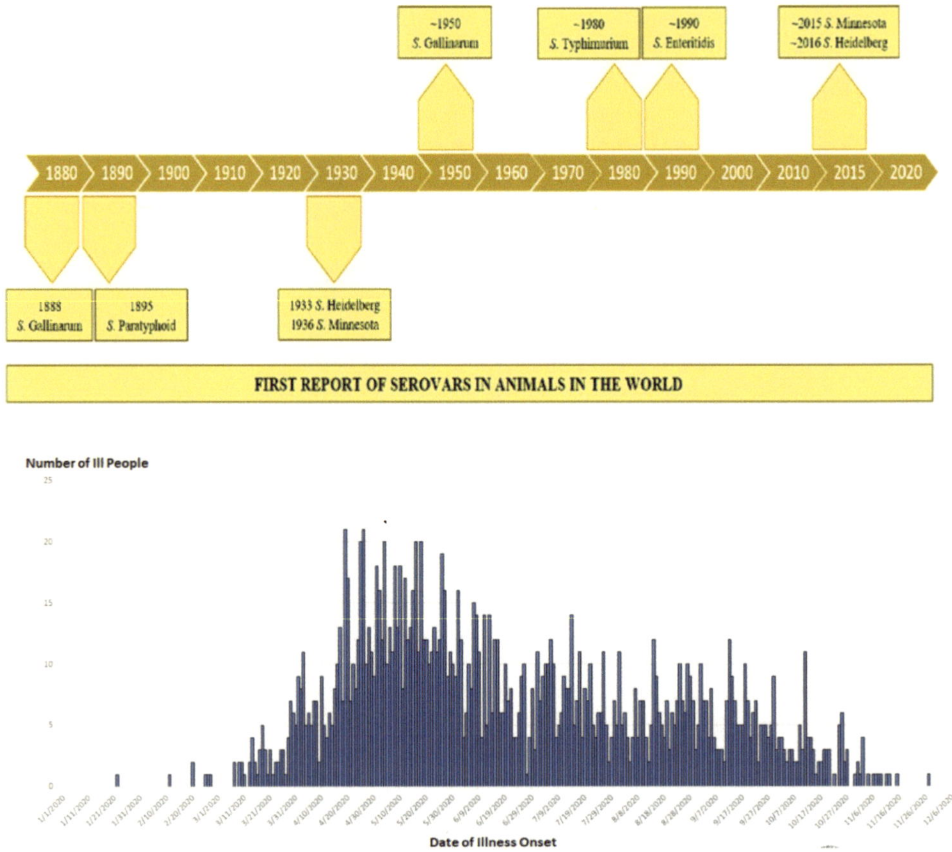

Figure 7.
Timeline of major salmonellosis outbreaks associated with birds [8, 50].

transmit zoonotic diseases through contaminated food. Furthermore, the emergence of antimicrobial-resistant strains of *Salmonella* in birds can complicate the treatment of *Salmonella* infections in both avian and human populations [46].

To summarize, *Salmonella* infections from avian sources pose substantial public health issues due to their potential to cause outbreaks, severe disease, and mortality. The presence of antimicrobial-resistant strains in birds adds complexity to the issue, requiring thorough monitoring, preventive, and control strategies to reduce the danger of transmission to people and other animal species (**Figure** 7).

10. Preventive measures

10.1 Strategies for reducing bird-related salmonellosis

Includes wildlife management, improving food hygiene, and public education [51]. Controlling *Salmonella* in chicken is paramount for public health, as it primarily contributes to human foodborne illnesses. It is crucial to employ efficient techniques to decrease the occurrence of salmonellosis caused by birds, hence minimizing the potential for transmission to humans and other animals. Various crucial strategies have been found to tackle this problem, such as implementing wildlife management, enhancing food cleanliness, and promoting public education.

10.2 Biosecurity practices in poultry farms

Critical for preventing the spread of *Salmonella* in poultry products [52]. Implementing biosecurity measures in poultry farms is crucial for mitigating the transmission of *Salmonella* in poultry products. Stringent biosecurity measures, such as thorough cleaning and disinfection, are crucial for minimizing disease occurrence in chicken farms and food processing facilities. Incorporating comprehensive hygiene and biosecurity measures into the whole poultry management strategy is crucial. The foremost step in preventing infections is the introduction of birds that are free from *Salmonella*. Furthermore, implementing limited entry, consistent surveillance of chicken populations, and prompt response upon identification of *Salmonella* are crucial elements of efficient biosecurity measures [53].

10.3 Non-antibiotic approaches

Administering postbiotics, which are non-antibiotic methods, has been proven to significantly decrease *Salmonella*-related infections in chickens. Using non-antibiotic options like probiotics, synbiotics, phytobiotics, and vaccinations has proven productive in managing *Salmonella* infections in laying hens, broilers, turkeys, and quails. It is crucial to use these non-antibiotic measures to avoid sickness and enhance the productive performance of birds while decreasing the dependence on antibiotics in poultry production [28].

10.4 Public education and awareness

Raising awareness about transmission risks and preventive measures is vital. Promoting an understanding of the hazards of transmitting salmonellosis through

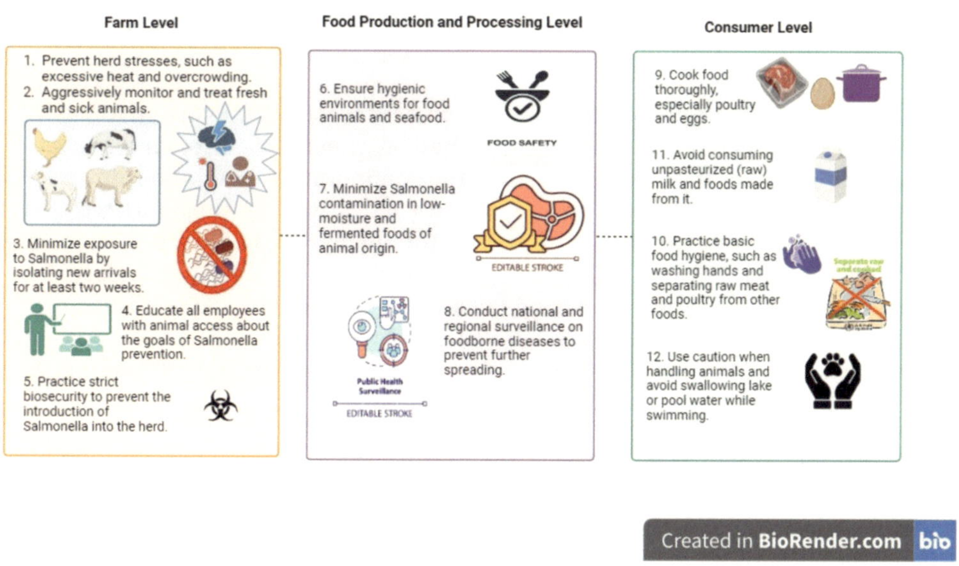

Figure 8.
Flowchart of the implementation of preventive measures at different levels.

birds and taking preventive steps to minimize these risks is crucial. Public education programs should prioritize promoting proper hygienic practices, including regularly cleaning feeders, ensuring various food sources, and monitoring bird behavior. Furthermore, it is crucial to educate poultry farmers and workers on the significance of implementing rigorous biosecurity protocols and exercising judicious practices when using antimicrobial agents. This is necessary to effectively curb the dissemination of *Salmonella* within chicken populations and mitigate the potential for foodborne transmission to people [9]. To summarize, it is crucial to employ all-encompassing approaches, such as efficient biosecurity measures, immunization, non-antibiotic substitutes, and public awareness campaigns, to diminish avian-associated salmonellosis and mitigate the likelihood of human transmission. These procedures must be implemented at every stage of chicken meat and egg production to guarantee the safety and excellence of poultry products and safeguard public health (**Figure 8**).

11. Research and future directions

11.1 Ongoing research in bird-associated *Salmonella*

Research focuses on understanding transmission dynamics, developing vaccines, and improving surveillance. Current investigations on bird-associated *Salmonella* have focused on understanding transmission dynamics, formulating vaccines, and enhancing surveillance. Research has been undertaken to ascertain the frequency and genetic makeup of *Salmonella* in wild birds across different areas. Research conducted in the western and southwestern regions of the United States discovered that the incidence of *Salmonella* in wild birds ranged from 0.5% to 6.5%. This study detected 24 avian species, with the eastern bluebird being the most prevalent [22]. The ability to detect *Salmonella* in wild birds to acquire and spread the pathogen has emphasized the significance of monitoring and managing efforts.

11.2 Innovations and technologies

Technological advancements in detection and control measures are ongoing. Continuous progress is being made in the technology field to identify and manage *Salmonella*. An area of innovation is the advancement of non-antibiotic approaches to manage *Salmonella* illness in chickens. Empirical research has substantiated that ingesting postbiotics through the mouth substantially diminishes *Salmonella*-related illnesses in chickens [28]. This provides a promising non-antibiotic option for reducing *Salmonella* infections in laying hens, broilers, turkeys, and quails facing challenges. Furthermore, progress in omics sciences has facilitated a more profound comprehension of *Salmonella's* genetic and molecular attributes, thereby aiding in the creation of inventive control strategies.

11.3 Challenges and opportunities

The evolving nature of *Salmonella* strains and changing environmental conditions present both challenges and opportunities for research. The dynamic

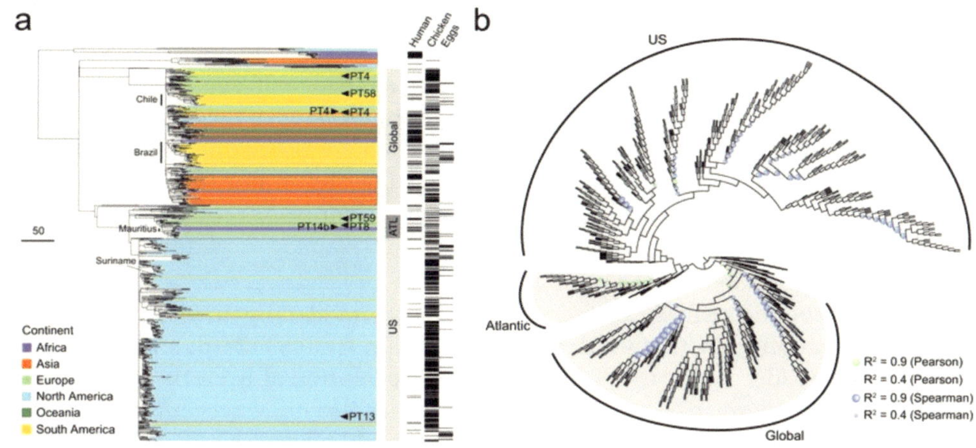

Figure 9.
Graphical representation of the research landscape of bird-associated salmonellosis. Maximum-likelihood phylogeny of 914 poultry, human, and other isolates from 46 countries. The tree is midpoint-rooted. There are three main lineages: Global, Atlantic, and United States. Each continent-originating isolate is color-coded. Human and poultry (chickens and eggs) isolates were included. In Chile, Brazil, and Mauritius poultry, Salmonella enteritidis *clades were identified. Isolates with publicly available phage types are arrow-headed. The Atlantic. The scale bar shows 50 SNPs. b circular cladogram of the maximum-likelihood phylogeny of 914 isolates. Internal nodes with a spearman or Pearson correlation between isolation years and branch lengths > 0.4 are shown by colored circles. The circle sizes are proportional to the R^2 values (0.4–0.9). Gray shading indicates high temporal SNP increases in the global and Atlantic lineages [54].*

characteristics of *Salmonella* strains and shifting environmental circumstances pose obstacles and prospects for scientific investigation. The occurrence of *Salmonella* in wild birds exhibits regional and temporal variability, with swings in prevalence observed throughout different years and geographic areas [22]. This fluctuation presents difficulty in monitoring and control endeavors. Nevertheless, it also provides a chance to obtain a more profound understanding of the elements that influence the incidence of *Salmonella* in wild bird populations. The rise of antimicrobial-resistant strains in avian species poses a substantial obstacle. Nevertheless, it has also stimulated the advancement of non-antibiotic approaches and the investigation of cutting-edge technology for managing *Salmonella* infections in poultry (**Figure 9**).

12. Conclusion

12.1 Summary of key findings

Birds play a significant role in the epidemiology of salmonellosis, serving as both reservoirs and vectors. The studies on the role of garden birds in the transmission of salmonellosis reveal a multifaceted issue that intersects wildlife ecology, public health, and environmental science. The research underscores the need for a holistic approach to managing this public health challenge, involving careful monitoring of bird populations, public education on safe bird interaction practices, and strategic wildlife management to mitigate the risks of salmonellosis transmission.

12.2 Extended analysis of key studies and findings

- Comprehensive view: this study was not limited to identifying garden birds as reservoirs. It also explored how human interaction with these birds, particularly in urban and suburban gardens, potentially increases the risk of salmonellosis transmission. The research suggested a more nuanced understanding of bird-human interactions and their implications for disease spread [4].

- Beyond genomic analysis: the genomic analysis provided by this study gave insights into the genetic similarities of *Salmonella* strains in birds and humans. This has broader implications for tracking the evolution of bacteria and understanding their spread across species. This finding also underscores the need for hygiene and safety in human interactions with wild birds [6].

- Specific bird species as reservoirs: by focusing on specific species like the Eurasian siskin and greenfinch, this study highlighted the varied susceptibility of different bird species to *Salmonella*. It paved the way for targeted public health advisories and wildlife management strategies, especially in regions where these species are common [55].

- Host resistance and poultry breeding: investigating host resistance at the genetic level opens up potential pathways for breeding disease-resistant poultry. This could have significant implications for the poultry industry and food safety, reducing the risk of salmonellosis transmission from poultry to humans [49].

- Seasonal dynamics and disease monitoring: understanding the seasonal patterns of *Salmonella* in wild birds such as greenfinches provides critical information for timely surveillance and intervention strategies. This research is vital for planning public health initiatives and wildlife conservation efforts [56].

- Risks of supplementary feeding: the study on supplementary feeding practices sheds light on how human activities can inadvertently increase disease transmission risks. It emphasizes the need for responsible wildlife feeding and the design of bird feeders that minimize the spread of pathogens [5].

- Diversity of *Salmonella* strains: the discovery of host-adapted *Salmonella* phage types in British garden birds suggested that the interaction between *Salmonella* and its bird hosts is more complex than previously thought. This finding is crucial for understanding the disease ecology and for developing more effective control measures [1].

- Environmental contamination: highlighting the risk of environmental contamination at public events like agricultural fairs, this study illustrates how *Salmonella* can spread in communal settings. This finding underscores the importance of sanitation and public awareness to prevent outbreaks [57].

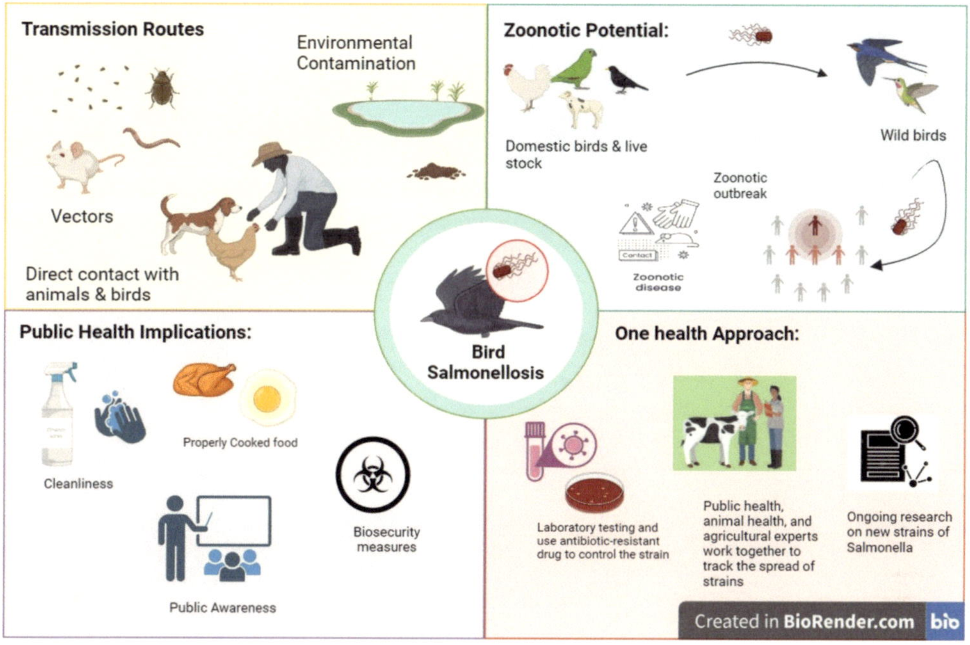

Figure 10.
Conceptual diagram summarizing the interconnectivity of the chapter's key points.

12.3 The ongoing importance of studying birds in salmonellosis

Consistent surveillance and investigation are crucial for effectively addressing the public health hazards linked to avian salmonellosis. The continued significance of researching salmonellosis in birds lies in their capacity to acquire and disseminate the bacterium, which presents a potential threat to human well-being. Human sickness outbreaks have been linked to direct interactions with untamed avian species. Moreover, wild birds can act as reservoirs for zoonotic infections, such as *Salmonella*. The continued significance of investigating avian involvement in salmonellosis is in the capacity of these birds to acquire and disseminate the bacterium, hence constituting a potential hazard to human well-being. Continuous surveillance and research are crucial for effectively managing the public health concerns linked to bird-associated salmonellosis. These efforts aid in assessing the extent of the danger of *Salmonella* transmission from wild birds to humans and in devising efficient ways for monitoring and controlling the spread of the disease (**Figure 10**).

Author details

Amina Tufail[1*] and Yusra Ashfaq[2]

1 Institute of Zoology, Chinese Academy of Sciences, Beijing, China

2 Institute of Zoology, University of the Punjab, Lahore, Pakistan

*Address all correspondence to: aminatufail25@gmail.com

References

[1] Lawson B et al. Pulsed-field gel electrophoresis supports the presence of host-adapted *Salmonella enterica* subsp. enterica serovar *Typhimurium* strains in the British garden bird population. Applied and Environmental Microbiology. 2011;**77**(22):8139-8144

[2] Tizard I. Salmonellosis in wild birds. In: Seminars in Avian and Exotic Pet Medicine. Vol. 13, no. 2. WB Saunders; 2004. pp. 60-66

[3] Hall AJ, Saito EK. Avian wildlife mortality events due to salmonellosis in the United States, 1985-2004. Journal of Wildlife Diseases. 2008;**44**(3):585-593

[4] Lawson B et al. Epidemiological evidence that garden birds are a source of human salmonellosis in England and Wales. PLoS One. 2014;**9**(2):e88968

[5] Lawson B et al. Health hazards to wild birds and risk factors associated with anthropogenic food provisioning. Philosophical Transactions of the Royal Society B: Biological Sciences. 2018;**373**(1745):20170091

[6] Mather AE et al. Genomic analysis of *Salmonella enterica* serovar Typhimurium from wild passerines in England and Wales. Applied and Environmental Microbiology. 2016;**82**(22):6728-6735

[7] Krawiec M et al. Prevalence and genetic characteristics of *Salmonella* in free-living birds in Poland. BMC Veterinary Research. 2015;**11**:15

[8] Kipper D et al. Emergence, dissemination and antimicrobial resistance of the main poultry-associated *Salmonella* serovars in Brazil. Veterinary Sciences. 2022;**9**(8):405

[9] Andres VM, Davies RH. Biosecurity measures to control *Salmonella* and other infectious agents in pig farms: A review. Comprehensive Reviews in Food Science and Food Safety. 2015;**14**(4):317-335

[10] Aung KT et al. *Salmonella* in retail food and wild birds in Singapore—Prevalence, antimicrobial resistance, and sequence types. International Journal of Environmental Research and Public Health. 2019;**16**(21):4235

[11] Kapperud G, Stenwig H, Lassen J. Epidemiology of *Salmonella typhimurium* O: 4-12 infection in Norway: Evidence of transmission from an avian wildlife reservoir. American Journal of Epidemiology. 1998;**147**(8):774-782

[12] Ellis EM. *Salmonella* reservoirs in animals and feeds. Journal of the American Oil Chemists' Society. 1969;**46**(5):227-229

[13] Zeynep Ş et al. Distribution of *Salmonella* serovars of animal origin in Türkiye between 2015 and 2020. Etlik Veteriner Mikrobiyoloji Dergisi. 2015;**33**(2):7-14

[14] Martín-Maldonado B et al. Urban birds: An important source of antimicrobial resistant *Salmonella* strains in Central Spain. Comparative Immunology, Microbiology and Infectious Diseases. 2020;**72**:101519

[15] Fu Y et al. Low occurrence of multi-antimicrobial and heavy metal resistance in *Salmonella enterica* from wild birds in the United States. Environmental Microbiology. 2022;**24**(3):1380-1394

[16] Wales A, Lawes J. JMM Profile: *Salmonella enterica* serovar Gallinarum, biovars Pullorum and Gallinarum. Journal of Medical Microbiology. 2023;**72**(2):001653

[17] Smith HG et al. Salmonella enterica serovar Hvittingfoss in bar-tailed godwits (*Limosa lapponica*) from Roebuck Bay, Northwestern Australia. Applied and Environmental Microbiology. 2020;**86**(19):e01312-e01320

[18] Guard-Petter J. The chicken, the egg and *Salmonella enteritidis*. Environmental Microbiology. 2001;**3**(7):421-430

[19] Oosterom J. Epidemiological studies and proposed preventive measures in the fight against human salmonellosis. International Journal of Food Microbiology. 1991;**12**(1):41-51

[20] Ferrari RG et al. Worldwide epidemiology of *Salmonella* Serovars in animal-based foods: A meta-analysis. Applied and Environmental Microbiology. 2019;**85**(14):e00591-19

[21] Hoelzer K, Moreno Switt AI, Wiedmann M. Animal contact as a source of human non-typhoidal salmonellosis. Veterinary Research. 2011;**42**(1):1-28

[22] Smith JC, Varriano S, Dunn LL, Snyder WE, Shariat NW. Prevalence and molecular characterization of *Salmonella* isolated from wild birds in fresh produce environments. Frontiers in Microbiology. 2023;**14**:1272916. DOI: 10.3389/fmicb.2023.1272916

[23] Mead G, Barrow P. *Salmonella* control in poultry by 'competitive exclusion' or immunization. Letters in Applied Microbiology. 1990;**10**(6):221-227

[24] Cardoso MD et al. *Salmonella* spp. profiles isolated from seabird samples from the Brazilian coast. Preventive Veterinary Medicine. 2021;**193**:105413

[25] Tardone R et al. *Salmonella* in raptors and aquatic wild birds in Chile. Journal of Wildlife Diseases. 2020;**56**(3):707-712

[26] Minette H. Salmonellosis in the marine environment. A review and commentary. International Journal of Zoonoses. 1986;**13**(2):71-75

[27] Zhou X et al. A global dataset for prevalence of *Salmonella gallinarum* between 1945 and 2021. Scientific Data. 2022;**9**(1):495

[28] Ruvalcaba-Gómez JM, Villagrán Z, Valdez-Alarcón JJ, Martínez-Núñez M, Gomez-Godínez LJ, Ruesga-Gutiérrez E, et al. Non-antibiotics strategies to control *Salmonella* infection in poultry. Animals. 2022;**12**(1):102

[29] Zhang-Barber L, Turner A, Barrow P. Vaccination for control of *Salmonella* in poultry. Vaccine. 1999;**17**(20-21):2538-2545

[30] Vandeplas S et al. *Salmonella* in chicken: Current and developing strategies to reduce contamination at farm level. Journal of Food Protection. 2010;**73**(4):774-785

[31] Van Immerseel F et al. Feed additives to control *Salmonella* in poultry. World's Poultry Science Journal. 2002;**58**(4):501-513

[32] Butucel E et al. Farm biosecurity measures and interventions with an impact on bacterial biofilms. Agriculture. 2022;**12**(8):1251

[33] Silva C, Calva E, Maloy S. One health and food-borne disease: *Salmonella* transmission between humans, animals, and plants. Microbiology Spectrum. 2014;**2**(1) p. 2.1. 08

[34] Christidis T et al. A comparative exposure assessment of foodborne, animal contact and waterborne transmission routes of *Salmonella* in Canada. Food Control. 2020;**109**:106899

[35] Liu H, Whitehouse CA, Li B. Presence and persistence of *Salmonella* in water: The impact on microbial quality of water and food safety. Frontiers in Public Health. 2018;**6**:159

[36] Liu B et al. The vertical transmission of *Salmonella enteritidis* in a one-health context. One Health. 2023;**16**:100469

[37] Lacroix-Lamandé S et al. Differential *Salmonella typhimurium* intracellular replication and host cell responses in caecal and ileal organoids derived from chicken. Veterinary Research. 2023;**54**(1):63

[38] Bäumler AJ, Hargis BM, Tsolis RM. Tracing the origins of *Salmonella* outbreaks. Science. 2000;**287**(5450):50-52

[39] Fu Y et al. *Salmonella enterica* serovar *typhimurium* isolates from wild birds in the United States represent distinct lineages defined by bird type. Applied and Environmental Microbiology. 2022;**88**(6):e0197921

[40] Pye HV et al. Strain and serovar variants of *Salmonella enterica* exhibit diverse tolerance to food chain-related stress. Food Microbiology. 2023;**112**:104237

[41] Rodríguez-Hernández R et al. Molecular characterization of *Salmonella* spp. isolates from wild Colombian babilla (*Caiman crocodilus fuscus*) isolated in situ. Animals. 2022;**12**(23):3359

[42] Castro-Vargas RE et al. Antibiotic resistance in *Salmonella* spp. isolated from poultry: A global overview. Veterinary World. 2020;**13**(10):2070-2084

[43] Cosby DE et al. *Salmonella* and antimicrobial resistance in broilers: A review. Journal of Applied Poultry Research. 2015;**24**(3):408-426

[44] Das T et al. Antimicrobial resistance profiling and burden of resistance genes in zoonotic *Salmonella* isolated from broiler chicken. Veterinary Medicine and Science. 2022;**8**(1):237-244. DOI: 10.1002/vms3.648 Available from: https://www.ncbi.nlm.nih.gov/pmc/articles/PMC8788989/

[45] Card RM et al. Multidrug-resistant non-typhoidal Salmonella of public health significance recovered from migratory birds in Bangladesh. Frontiers in Microbiology. 2023;**14**:1162657

[46] Lutful Kabir SM. Avian colibacillosis and salmonellosis: A closer look at epidemiology, pathogenesis, diagnosis, control and public health concerns. International Journal of Environmental Research and Public Health. 2010;**7**(1):89-114

[47] Patel K et al. Human salmonellosis outbreak linked to *Salmonella typhimurium* epidemic in wild songbirds, United States, 2020-2021. Emerging Infectious Diseases. 2023;**29**(11):2298

[48] Centers for Disease Control and Prevention. Salmonella: Investigation Details. National Center for Emerging and Zoonotic Infectious Diseases (NCEZID), Division of Foodborne, Waterborne, and Environmental Diseases (DFWED); 2023. Available from: https://www.cdc.gov/salmonella/backyardpoultry-05-23/details.html

[49] Sadeyen J-R et al. *Salmonella* carrier-state in hens: Study of host resistance by a gene expression approach. Microbes and Infection. 2006;**8**(5):1308-1314

[50] Centers for Disease Control and Prevention. Salmonella. National Center for Emerging and Zoonotic Infectious Diseases (NCEZID), Division of Foodborne, Waterborne, and

Environmental Diseases (DFWED); 2020. Available from: https://www.cdc.gov/salmonella/backyardpoultry-05-20/epi.html

[51] Laptev GY et al. Effects of essential oils-based supplement and *Salmonella* infection on gene expression, blood parameters, cecal microbiome, and egg production in laying hens. Animals. 2021;**11**(2):360

[52] Parsons BN et al. Invasive non-typhoidal *Salmonella typhimurium* ST313 are not host-restricted and have an invasive phenotype in experimentally infected chickens. PLoS Neglected Tropical Diseases. 2013;7(10):e2487

[53] Zamora-Sanabria R, Molina A. Preharvest Salmonella Risk Contamination and the Control Strategies. In: Current Topics in Salmonella and Salmonellosis. 2017. DOI: 10.5772/67399

[54] Li S, He Y, Mann DA, Deng X. Global spread of Salmonella Enteritidis via centralized sourcing and international trade of poultry breeding stocks. Nature Communications. 2021;**12**(1):1-12. DOI: 10.1038/s41467-021-25319-7. Available from: https://www.nature.com/articles/s41467-021-25319-7

[55] Krawiec M, Pietkiewicz M, Wieliczko A. *Salmonella* spp. as a cause of mortality and clinical symptoms in free-living garden bird species in Poland. Polish Journal of Veterinary Sciences. 2014;**17**(4):729-731. DOI: 10.2478/pjvs-2014-0107

[56] Grant D, Todd PA, Pennycott T. Monitoring wild greenfinch (*Carduelis chloris*) for *Salmonella enterica* typhimurium. Ecological Research. 2007;**22**(4):571-574

[57] Pabilonia K et al. Environmental *Salmonella* in agricultural fair poultry exhibits in Colorado. Zoonoses and Public Health. 2014;**61**(2):138-144

Chapter 3

Accurate Identification of *Salmonella enterica* in Calves

Abdul Kabir, Momin Khan and Anees Ur Rahman

Abstract

Salmonella enterica is a bacterium that can cause serious infections in humans and animals, especially cattle. The identification and differentiation of *S. enterica* serotypes in cattle is important for epidemiological surveillance, disease prevention and control, and public health protection. However, the current methods and techniques for *S. enterica* detection have various challenges and limitations, such as low sensitivity and specificity, high cost and complexity, and the need for viable and pure bacterial cultures. Therefore, there is a need for further research and development of novel and improved methods and techniques that can overcome these challenges and provide reliable and accurate information on *S. enterica* serotypes in cattle. Such information can help to improve the understanding of the epidemiology, pathogenesis, and transmission of *S. enterica* in cattle, as well as to design and implement effective strategies for its prevention and control. This chapter reviews the current methods and techniques for *S. enterica* detection, such as culture-based methods, biochemical methods, molecular methods, phage-based methods, and biosensor methods, and discusses their advantages and disadvantages, as well as their future trends and perspectives.

Keywords: *Salmonella enterica*, cattle, detection, identification, serotypes, methods, techniques

1. Introduction

Salmonella is a bacterium that can cause serious infections in humans and animals, especially in young, immunocompromised, or stressed individuals [1]. It has over 2000 strains, which are classified into serotypes based on their antigenic properties [2]. The serotypes differ in their pathogenicity and prevalence in different host species, including cattle. Some serotypes, such as *S. Typhimurium* and *S. Enteritidis*, are able to cause systemic disease in a wide range of hosts, while others, such as *S. Dublin* and *S. Choleraesuis*, are more host-specific and adaptable to cattle and pigs, respectively [2]. The prevalence of *Salmonella* serotypes in cattle varies by region, production system, and sampling method. A systematic review and meta-analysis of studies published between 2000 and 2017 estimated a pooled prevalence of *Salmonella* in apparently healthy cattle of 9%, with the highest prevalence in North America (16%) and the lowest in Europe (2%) [3]. The most frequently reported serotypes in cattle are *S. Montevideo, S. Typhimurium, S. Kentucky, S. Meleagridis,* and *S. Anatum* [3]. However, the serotype distribution may change over time due to factors such as antimicrobial resistance, vaccination, and animal trade [4, 5]. In cattle, *Salmonella enterica* can cause salmonellosis, a disease characterized by fever, diarrhea, dehydration, septicemia, abortion, and sometimes death [2]. The incidence and impact of salmonellosis in cattle vary depending on the serotype of *S. enterica*, the host susceptibility, and environmental factors [3–5]. Therefore, accurate identification and differentiation of *S. enterica* serotypes in cattle is essential for epidemiological surveillance, disease prevention and control, and public health protection [6–8].

The conventional methods for identifying and differentiating *S. enterica* serotypes in cattle are based on bacterial culture and biochemical and serological tests. However, these methods are time-consuming, labor-intensive, and may not be able to detect all the serotypes or distinguish closely related serotypes [9]. Moreover, these methods require viable and pure bacterial cultures, which may not always be available or feasible. Molecular methods can be used to identify and differentiate *S. enterica* serotypes in cattle directly from clinical samples, such as feces, tissues, or fluids [5, 10, 11]. These methods are based on the detection and amplification of specific DNA sequences of *S. enterica*, such as the invA gene, which encodes an invasion protein, or the rfb gene cluster, which encodes the O antigen [12]. Molecular methods can also provide information on the genetic diversity, phylogeny, and epidemiology of *S. enterica* serotypes in cattle [13].

In this chapter, we will review the current molecular methods for identifying and differentiating *S. enterica* serotypes in cattle, such as polymerase chain reaction (PCR), multiplex PCR, real-time PCR, loop-mediated isothermal amplification (LAMP), and whole-genome sequencing (WGS) [12–17]. We will also discuss the advantages and limitations of each method, as well as the future perspectives and challenges for improving the accuracy and reliability of *S. enterica* identification and differentiation in cattle. Furthermore, for the first time, we will present a case report of a new serovar of *S. enterica* that was detected in Mediterranean buffalo calves (*Bubalus bubalis*) using a combination of molecular and conventional methods. This case illustrates the importance and potential of molecular methods for discovering and characterizing novel *S. enterica* serotypes in cattle and other animal species.

1.1 Some application examples of the molecular methods for identifying *Salmonella* in cattle include the following

PCR: PCR is a technique that amplifies a specific DNA fragment from a complex mixture of DNA. PCR can be used to detect *S. enterica* in cattle by targeting the invA gene, which is present in all *S. enterica* serotypes. PCR can also be used to differentiate *S. enterica* serotypes by targeting the rfb gene cluster, which encodes the O antigen [12]. For example, a study by Eriksson et al. compared the performance of PCR, enzyme-linked immunosorbent assay (ELISA), and culture methods for *Salmonella* detection in fecal samples from cattle, pigs, and poultry. The results showed that PCR had a higher sensitivity and specificity than ELISA and culture methods and could detect *S. enterica* serotypes that were not detected by the other methods [13].

Multiplex PCR: Multiplex PCR is a technique that amplifies multiple DNA fragments simultaneously in a single reaction. Multiplex PCR can be used to identify and differentiate *S. enterica* serotypes in cattle by targeting multiple genes or regions that are specific for different serotypes [12]. A study by D'Angelo et al. used multiplex

PCR to identify and differentiate two *S. enterica* serotypes, *S. enterica* subsp. *enterica* O:35 and a new serovar of *S. enterica*, in fecal samples from water buffalo calves that showed severe gastroenteritis. Multiplex PCR targeted the invA gene and the rfb gene cluster and could distinguish the two serotypes based on the size of the amplified products [14].

Real-time PCR: Real-time PCR is a technique that monitors the amplification of DNA in real time by using fluorescent probes or dyes. Real-time PCR can be used to quantify *S. enterica* in cattle by measuring the fluorescence intensity of the amplified products. Real-time PCR can also be used to differentiate *S. enterica* serotypes in cattle by using specific probes or melting curve analysis [15]. A study by Harbottle et al. used real-time PCR to compare the genetic diversity of *S. enterica* serotype Newport isolates from cattle, humans, and food. Real-time PCR was used to target the invA gene, and melting curve analysis was used to differentiate the isolates based on single nucleotide polymorphisms (SNPs) in the invA gene [15].

LAMP: LAMP is a technique that amplifies DNA under isothermal conditions by using a set of primers and a polymerase with strand displacement activity. LAMP can be used to detect *S. enterica* in cattle by targeting the invA gene or other genes that are specific to *S. enterica* [18]. LAMP can also be used to differentiate *S. enterica* serotypes in cattle by using different sets of primers or adding loop primers that are specific for different serotypes. For example, a study by Wang et al. used LAMP to detect and differentiate *S. enterica* serotypes Typhimurium, Enteritidis, and Dublin in milk samples. The LAMP assay targeted the invA gene and used loop primers that were specific for the three serotypes. The LAMP assay could detect and differentiate the three serotypes within 60 min, with a sensitivity of 10 CFU/mL [16, 17].

WGS: WGS is a technique that sequences the entire genome of an organism. This study analyzed the WGS data of *S. enterica* serotypes in cattle by comparing the genomic sequences of different isolates and identifying the genes or regions that are unique or variable for different serotypes [13]. WGS can also provide information on the phylogeny, evolution, and epidemiology of *S. enterica* serotypes in cattle [14, 15]. For example, a study by EFSA (European Food Safety Authority) used WGS to investigate the public health risks related to the consumption of raw drinking milk. The study analyzed the WGS data of *S. enterica* isolates from raw milk, cheese, and humans and identified the serotypes, subtypes, and antimicrobial resistance profiles of the isolates. This study also traced the sources and transmission routes of *S. enterica* in raw milk and cheese production [19–21].

1.2 The case report of a new serovar of *S. enterica* in Mediterranean buffalo calves is as follows

The case report of the new serovar of *S. enterica* in Mediterranean buffalo calves is based on the analysis of fecal samples collected from live water buffalo calves that showed signs of severe gastroenteritis. The samples were cultured on selective media and incubated at 37°C for 24 hours. The colonies were identified by biochemical tests and serotyping using specific antisera. The molecular methods included PCR amplification of the invA gene and the 16S rRNA gene, followed by sequencing and phylogenetic analysis. The results confirmed the presence of two *S. enterica* serotypes, *S. enterica O:35* and a new serovar of *S. enterica*, in the intestinal contents of the water buffalo calves. The new serovar of *S. enterica* was named *S. enterica* subsp. *enterica* serovar Buffalo [22].

2. Sample collection, culture methods, and biochemical tests used for *S. enterica*

The conventional methods for the identification and differentiation of *S. enterica* serotype in cattle are based on the culture and biochemical characterization of the isolates obtained from the samples. The culture methods involved the enrichment, isolation, and confirmation of *S. enterica* from various types of samples, such as feces, milk, tissues, or environmental swabs. Biochemical tests are used to determine the serogroup and serotype of the isolates based on their antigenic properties, such as the presence of O (somatic), H (flagellar), and Vi (capsular) antigens [23, 24]. These methods are widely used in veterinary diagnostic laboratories and epidemiological studies, as they are relatively simple, inexpensive, and standardized.

In this section, we will describe the conventional methods for sample collection, culture, and biochemical identification of *S. enterica* in cattle, as well as their advantages and disadvantages. We will also mention some of the challenges and limitations of these methods and suggest some possible solutions and improvements.

2.1 Sample collection

The first step in the detection of *S. enterica* in cattle is the collection of appropriate samples [18, 25, 26]. The types of samples that can be collected for *S. enterica* detection include feces, tissues, fluids, food, and environmental samples [22]. The choice of sample depends on the purpose and objective of the testing, the clinical signs and status of the animals, and the availability and feasibility of the sampling methods [27].

Fecal samples are the most common and convenient samples for *S. enterica* detection in cattle because they can reflect intestinal colonization and shedding of bacteria [28]. Fecal samples can be collected from individual animals or from pooled samples of a group of animals. Fecal samples can be collected by rectal swabs, fecal grab, or fecal cup. Rectal swabs are easy and quick to perform, but they may not collect enough material and may be contaminated by the rectal mucosa [29]. Fecal grab is the method of collecting a handful of feces from the rectum of the animal, which can provide more material and less contamination, but it may be difficult and stressful for the animal and the sampler. A fecal cup is a method of collecting feces from the ground after the animal defecates, which can avoid the stress and discomfort of the animal and the sampler, but it may be affected by environmental contamination and degradation [30].

Tissue samples are another type of sample that can be collected for *S. enterica* detection in cattle, especially in cases of systemic infection, abortion, or postmortem examination [31]. Tissue samples can include lymph nodes, spleen, liver, kidney, lung, brain, placenta, fetus, and other organs that may be affected by *S. enterica* [18, 32, 33]. Tissue samples can be collected by biopsy, necropsy, or abortion investigation. A biopsy is a method of collecting a small piece of tissue from a living animal, which can provide early diagnosis and treatment, but it may be invasive and risky for the animal and the sampler [8]. Necropsy is the method of collecting tissue samples from a dead animal, which can provide comprehensive and conclusive information, but it may be too late and costly for the prevention and control of the disease. Abortion investigation is the method of collecting tissue samples from the aborted fetus and placenta, which can provide valuable information on the cause and source of the abortion, but it may be difficult and sensitive to perform [34].

Fluid samples are another type of sample that can be collected for *S. enterica* detection in cattle, especially in cases of septicemia, mastitis, or urinary tract infection. Fluid samples can include blood, milk, urine, and other body fluids that may be infected by *S. enterica* [35]. Fluid samples can be collected by venipuncture, milking, catheterization, or aspiration. Venipuncture is the method of collecting blood from a vein of the animal, which can provide information on the systemic infection and immune response, but it may be stressful and painful for the animal and the sampler [36, 37]. Milking is the method of collecting milk from the udder of the animal, which can provide information on mastitis and udder health, but it may be influenced by the milking hygiene and frequency [38]. Catheterization is the method of collecting urine from the bladder of the animal, which can provide information on the urinary tract infection and kidney function, but it may be invasive and risky for the animal and the sampler [32, 39–41]. Aspiration is the method of collecting fluid from a cavity or lesion of the animal, which can provide information on the local infection and

Sampling method	Advantages	Disadvantages
Venipuncture	Provides information on the systemic infection and immune response • Can be performed on any animal with a visible vein • Can be done with minimal equipment and training	Stressful and painful for the animal and the sampler • May cause bleeding, bruising, or infection at the puncture site • May be difficult to obtain enough blood volume or quality
Milking	• Provides information on the mastitis and udder health • Can be done on any lactating animal • Can be integrated with the routine milking procedure	• Influenced by the milking hygiene and frequency • May cause contamination or cross-contamination of the milk samples • May be affected by the stage of lactation or the presence of antibiotics
Catheterization	• Provides information on the urinary tract infection and kidney function • Can provide a sterile and uncontaminated urine sample • Can be done on any animal with a bladder	• Invasive and risky for the animal and the sampler • May cause injury, infection, or inflammation of the urinary tract • May require sedation or anesthesia of the animal
Aspiration	• Provides information on the local infection and inflammation • Can provide a direct and concentrated fluid sample • Can be done on any animal with a cavity or lesion	• Difficult and complicated to perform • May cause damage, bleeding, or infection of the tissue or organ • May require specialized equipment and training
Venipuncture	• Provides information on the systemic infection and immune response • Can be performed on any animal with a visible vein • Can be done with minimal equipment and training	• Stressful and painful for the animal and the sampler • May cause bleeding, bruising, or infection at the puncture site • May be difficult to obtain enough blood volume or quality

Table 1.
Advantages and disadvantages of fluid sampling methods for S. enterica *detection in cattle.*

inflammation, but it may be difficult and complicated to perform. The advantages and disadvantages of each sampling method are summarized in **Table 1**.

The quality and quantity of the fluid samples collected for *S. enterica* detection in cattle depend on various factors, such as the sampling time, location, method, storage, and transport [42]. The sampling time should be chosen according to the epidemiological situation, clinical signs, and shedding patterns of the animals [43]. The sampling location was selected according to the source and route of the infection, the distribution and concentration of the bacteria, and the accessibility and availability of the sampling sites. The sampling method should be chosen according to the type and nature of the fluid, the purpose and objective of the testing, and the feasibility and acceptability of the sampling procedure. The storage and transport of fluid samples should be performed according to the temperature and humidity conditions, the preservation and protection of the samples, and the duration and distance of the storage and transport [34, 44].

Some general guidelines and recommendations for optimal fluid sample collection and handling for *S. enterica* detection in cattle are as follows: Samples are collected as soon as possible after the onset of the infection or the outbreak or before the initiation of the treatment or the intervention. Samples from multiple animals or sources or from multiple sites or locations were collected to increase the probability and representativeness of detection [45]. The samples were collected using sterile and appropriate equipment and containers and labeled clearly and correctly. Store and transport samples at refrigerated temperatures (4°C) and in insulated and sealed containers. Freezing and thawing cycles, exposure to sunlight or heat, and prolonged storage or transport times were avoided. The samples were processed and analyzed as soon as possible after arrival at the laboratory or stored at frozen temperatures (−20°C or lower) until further analysis [46].

The current standards for sampling quantities of different fluid samples for *S. enterica* detection in cattle vary depending on the type of fluid, the method of analysis, and the level of sensitivity and specificity required. However, some general recommendations are as follows: collect at least 10 ml of blood per animal, use a sterile syringe and needle, and transfer it to a sterile tube containing an anticoagulant, such as ethylenediaminetetraacetic acid (EDTA) or heparin [35, 47, 48]. At least 50 ml of milk per animal was collected using a sterile container and mixed well before aliquoting [36]. At least 10 ml of urine per animal was collected using a sterile catheter and syringe and transferred to a sterile tube or container [37, 49]. At least 5 ml of fluid per animal was collected using a sterile needle and syringe and transferred to a sterile tube or container [38]. The sampling quantities of different fluid samples are summarized in **Table 2**.

Food samples are another type of sample that can be collected for *S. enterica* detection in cattle, especially in cases of foodborne outbreaks or surveillance. Food samples can include raw or processed meat, milk, cheese, and other products that

Fluid sample	Sampling quantity
Blood	10 ml
Milk	50 ml
Urine	10 ml
Fluid	5 ml

Table 2.
Sampling quantities of different fluid samples for Salmonella enterica *detection in cattle.*

Sampling method	Advantages	Disadvantages
Sampling plans	• Provides representative and consistent information • Can be designed according to the specific criteria and objectives • Can be validated and standardized	• Complex and costly to implement • May require large number or size of samples • May not detect rare or sporadic contamination
Random sampling	• Provides unbiased and general information • Can be applied to any population or lot • Can be simple and easy to perform	• Not sufficient or efficient to detect low levels of contamination • May not reflect the variability or heterogeneity of the population or lot • May not identify the source or cause of the contamination
Targeted sampling	• Provides focused and relevant information • Can be directed to the suspected or known sources or locations • Can be sensitive and specific to detect contamination	• Not comprehensive or objective to assess the overall situation • May miss other potential sources or locations • May be influenced by the prior knowledge or bias

Table 3.
Advantages and disadvantages of food sampling methods for Salmonella enterica *detection in cattle.*

may be contaminated by *S. enterica*. Food samples can be collected by sampling plans, random sampling, or targeted sampling. The advantages and disadvantages of each sampling method are summarized in **Table 3**.

Environmental samples are another type of sample that can be collected for *S. enterica* detection in cattle, especially in cases of environmental monitoring or investigation [31, 50, 51]. Environmental samples can include water, soil, bedding, feed, equipment, and other materials that may be contaminated by *S. enterica* [33]. Environmental samples can be collected by swabs, filters, scoops, or wipes. The advantages and disadvantages of each sampling method are summarized in **Table 4**.

The quality and quantity of food and environmental samples collected for *S. enterica* detection in cattle depend on various factors, such as the sampling time, location, method, storage, and transport [25, 52]. The sampling time should be chosen according to the epidemiological situation, clinical signs, and shedding patterns of the animals [28, 53]. The sampling location was selected according to the source and route of the infection, the distribution and concentration of the bacteria, and the accessibility and availability of the sampling sites. The sampling method should be chosen according to the type and nature of the sample, the purpose and objective of the testing, and the feasibility and acceptability of the sampling procedure. The storage and transport of the samples should be done according to the temperature and humidity conditions, the preservation and protection of the samples, and the duration and distance of the storage and transport [34, 54].

Some general guidelines and recommendations for optimal food and environmental sample collection and handling for *S. enterica* detection in cattle are as follows: Samples are collected as soon as possible after the onset of the infection or the outbreak or before the initiation of the treatment or the intervention [55]. Samples from multiple animals or sources or from multiple sites or locations were collected to increase the probability and representativeness of the detection. The samples were

Sampling method	Advantages	Disadvantages
Swabs	• Simple and convenient to perform • Can be used for any surface or object • Can be stored and transported easily	• Not collect enough material • Affected by the moisture and texture of the surface or object • May require enrichment or concentration steps
Filters	• Quantitative and sensitive to detect contamination • Can be used for water or air samples • Can be processed and analyzed directly	• Complex and expensive to perform • May require specialized equipment and training • May be influenced by the volume or flow rate of the sample
Scoops	• Abundant and representative of the sample • Can be used for soil, feed, or other loose material • Can be mixed and homogenized easily	• Bulky and messy to handle • May require subsampling or dilution steps • May be affected by the environmental conditions or degradation
Wipes	• Large and uniform of the sample • Can be used for flat or smooth surfaces or objects • Can be stored and transported easily	• Not suitable for rough or irregular surfaces or objects • May require enrichment or concentration steps • May cause cross-contamination or interference

Table 4.
Advantages and disadvantages of environmental sampling methods for S. enterica *detection in cattle.*

collected using sterile and appropriate equipment and containers and labeled clearly and correctly. Store and transport samples at refrigerated temperatures (4°C) and in insulated and sealed containers [56]. Freezing and thawing cycles, exposure to sunlight or heat, and prolonged storage or transport times were avoided. The samples were processed and analyzed as soon as possible after arrival at the laboratory or stored at frozen temperatures (−20°C or lower) until further analysis.

The current standards for sampling quantities of different food and environmental samples for *S. enterica* detection in cattle vary depending on the type of sample, the method of analysis, and the level of sensitivity and specificity required. However, some general recommendations are as follows: at least 25 g of meat or cheese should be collected per sample using a sterile knife or scissors and transferred to a sterile bag or container [39, 55]. At least 100 ml of milk per sample was collected using a sterile container and mixed well before aliquoting [57]. At least 10 g of swab, filter, scoop, or wipe was collected per sample using a sterile device and transferred to a sterile bag or container [40, 58]. The sampling quantities of different food and environmental samples are summarized in **Table 5**.

Some general guidelines and recommendations for optimal sample collection and handling for *S. enterica* detection in cattle are as follows:

- To ensure timely and accurate detection, samples were collected as soon as possible after the onset of the infection or the outbreak or before the initiation of the treatment or the intervention.

Food or environmental sample	Sampling quantity
Meat or cheese	25 g
Milk	100 ml
Swab, filter, scoop, or wipe	10 g

Table 5.
Sampling quantities of different food and environmental samples for S. enterica *detection in cattle.*

- To increase the probability and representativeness of detection, samples were collected from multiple animals or sources or from multiple sites or locations [59, 60].

- To prevent contamination and degradation of the samples, appropriate and sterile equipment, containers, and materials were used, and standard operating procedures and good laboratory practices were followed.

- To facilitate identification and tracking of the samples, the samples were labeled and recorded with relevant information, such as the date, time, location, animal, type, and method of the collection.

- To maintain the viability and integrity of the samples, they were stored and transported at refrigerated temperatures (2–8°C) and in sealed and leak-proof containers to avoid freezing, thawing, or exposing them to direct sunlight or heat [40, 61].

- To expedite the analysis and reporting of the results, the samples should be delivered and processed as soon as possible, preferably within 24 hours, or appropriate preservatives or stabilizers should be used.

2.2 Culture methods

Culture methods are widely used for *S. enterica* detection in cattle, as they can provide viable and pure cultures of the bacteria, which can be further identified and characterized by biochemical and molecular methods [45, 62]. However, culture methods are also time-consuming and labor-intensive and may not be able to detect all the serotypes or distinguish closely related ones. Moreover, they require viable and pure bacterial cultures, which may not always be available or feasible [28, 63, 64]. Therefore, there is a need for rapid, sensitive, and specific molecular methods that can identify and differentiate *S. enterica* serotypes in cattle directly from clinical samples, such as feces, tissues, or fluids [40, 61, 62].

The application of culture methods for *S. enterica* detection in cattle has been demonstrated in various studies and reports, which have shown the prevalence, incidence, distribution, and diversity of *S. enterica* serotypes in cattle populations, as well as their association with clinical signs, risk factors, and transmission routes [28, 61, 63]. Some examples of these studies and reports are as follows.

A study by the author [64] investigated the prevalence and serotypes of *S. enterica* in dairy cattle in Ontario, Canada, using culture and serotyping methods. The study found that the overall prevalence of *S. enterica* was 2.8% and that the most common serotypes were Typhimurium, Newport, and Dublin. The study also found that the prevalence of *S. enterica* was higher in calves, heifers, and cows with diarrhea and that the main risk factors were herd size, housing type, and feed type.

A study by the author [50] compared the culture and PCR methods for *S. enterica* detection in fecal samples of beef cattle in Nebraska, USA. The study found that the culture method detected *S. enterica* in 6.3% of the samples, while the PCR method detected *S. enterica* in 12.5% of the samples. The study also found that the most common serotypes detected by culture were Anatum, Montevideo, and Mbandaka, and that the PCR method had a higher sensitivity and specificity than the culture method.

A report by the author [65] analyzed the *S. enterica* isolates from cattle submitted to the National Veterinary Services Laboratories in the United States from 2014 to 2018, using culture and serotyping methods. The report found that the most common serotypes isolated from cattle were Dublin, Newport, Cerro, and Typhimurium and that the serotype distribution varied by region, season, and age group. The report also found that some serotypes, such as Dublin and Newport, were more likely to cause systemic infections, while others, such as Typhimurium and Montevideo, were more likely to cause enteric infections.

These examples show the usefulness and limitations of culture methods for *S. enterica* detection in cattle and the need for further research and development of novel and improved methods that can overcome these challenges and provide reliable and accurate information on *S. enterica* serotypes in cattle. Such information can help to improve the understanding of the epidemiology, pathogenesis, and transmission of *S. enterica* in cattle, as well as to design and implement effective strategies for its prevention and control.

The general procedure for the culture of *S. enterica* from the samples involves two steps: enrichment and isolation. Enrichment is the step of incubating the samples in a liquid medium that favors the growth of *S. enterica* and inhibits the growth of other bacteria. Isolation is the step of transferring the enriched samples to a solid medium that allows the differentiation of *S. enterica* from other bacteria based on their colony morphology and color. There are various types of media that can be used for enrichment and isolation of *S. enterica,* and they can be classified into selective and differential media. Selective media are media that contain substances that inhibit the growth of certain bacteria, while allowing the growth of others [66]. Differential media are media that contain substances that change color or produce a reaction when metabolized by certain bacteria, while not affecting others. Some media can be both selective and differential, depending on their composition and function [66].

Some of the commonly used media for the enrichment of *S. enterica* are tetrathionate broth (TT), which is a selective medium that contains tetrathionate and inhibits the growth of most Gram-positive and some Gram-negative bacteria, while allowing the growth of *S. enterica* and some other enteric bacteria. It also contains bile salts, which inhibit the growth of non-enteric bacteria, and iodine, which inhibits the growth of *Proteus* spp. and some other bacteria that can produce hydrogen sulfide. TT is used as a primary enrichment medium for *S. enterica* detection in fecal, food, and environmental samples [67]. Rappaport-Vassiliadis broth (RV or RVS): This is a selective medium that contains magnesium chloride and malachite green, which inhibit the growth of most bacteria, while allowing the growth of *S. enterica* and some other enteric bacteria. RV is used as a secondary enrichment medium for *S. enterica* detection in fecal, food, and environmental samples [39, 68]. Selenite cystine broth (SC): This is a selective medium that contains sodium selenite and cystine, which inhibit the growth of most bacteria, while allowing the growth of *S. enterica* and some other enteric bacteria. SC is used as a primary or secondary enrichment medium for *S. enterica* detection in fecal, food, and environmental samples [69].

Some of the commonly used media for the isolation of *S. enterica* are xylose lysine deoxycholate agar (XLD) and Hektoen enteric agar (HE). These are selective and differential media that contain xylose, lysine, and deoxycholate, which inhibit the growth of most Gram-positive and some Gram-negative bacteria, while allowing the growth of *S. enterica* and some other enteric bacteria. It also contains phenol red, which changes color from red to yellow when xylose is fermented, and sodium thiosulfate and ferric ammonium citrate, which produce a black precipitate when hydrogen sulfide is produced [70]. XLD is used as a primary isolation medium for *S. enterica* detection in fecal, food, and environmental samples. On XLD, *S. enterica* colonies appear as red colonies with or without a black center, while other bacteria appear as yellow colonies with or without a black center or are inhibited [50]. Hektoen enteric agar (HE): This is a selective and differential medium that contains lactose, sucrose, and salicin, which inhibit the growth of most Gram-positive and some Gram-negative bacteria, while allowing the growth of *S. enterica* and some other enteric bacteria. It also contains bromothymol blue and acid fuchsin, which change color from green to yellow when lactose, sucrose, or salicin is fermented, and sodium thiosulfate and ferric ammonium citrate, which produce a black precipitate when hydrogen sulfide is produced. HE is used as a primary or secondary isolation medium for *S. enterica* detection in fecal, food, and environmental samples [42]. On HE, *S. enterica* colonies appear as green or blue-green colonies with or without a black center, while other bacteria appear as yellow or orange colonies with or without a black center or are inhibited.

Bismuth sulfite agar (BS): This is a selective and differential medium that contains bismuth sulfite, which inhibits the growth of most bacteria, while allowing the growth of *S. enterica* and some other enteric bacteria. It also contains brilliant green, which inhibits the growth of most Gram-positive and some Gram-negative bacteria, and sodium thiosulfate and ferric ammonium citrate, which produce a black precipitate when hydrogen sulfide is produced [70, 71]. BS is used as a primary or secondary isolation medium for *S. enterica* detection in fecal, food, and environmental samples. On BS, *S. enterica* colonies appear as black or brown colonies, while other bacteria appear as green or blue colonies or are inhibited [72].

The advantages of culture methods for *S. enterica* detection are that they are relatively simple, inexpensive, and widely available and that they can provide viable and pure cultures of the bacteria, which can be further identified and characterized by biochemical and molecular methods [45, 72]. The disadvantages of culture methods are that they are time-consuming, labor-intensive, and may not be able to detect all the serotypes or distinguish closely related ones. Moreover, they require viable and pure bacterial cultures, which may not always be available or feasible [71]. Therefore, there is a need for rapid, sensitive, and specific molecular methods that can identify and differentiate *S. enterica* serotypes in cattle directly from clinical samples, such as feces, tissues, or fluids.

3. Advances in recent techniques for the detection of *Salmonella*

Advances in recent techniques for the detection of *S. enterica* is a type of bacteria that causes food poisoning and typhoid fever in humans and animals. It can be transmitted through contaminated water or food and can cause diarrhea, fever, stomach cramps, nausea, vomiting, and sometimes blood in the stool. The detection and identification of *S. enterica* is crucial for the diagnosis, treatment, and prevention of the infection, as well as for the surveillance and control of the disease outbreak [5, 6]. Different techniques for *S. enterica* detection have different advantages and disadvantages, depending on their principles, procedures, performance, and applications. In this section, we will briefly compare some of the conventional and novel techniques for *S. enterica* detection in cattle, such as culture-based methods, molecular methods, immunological methods, phage-based methods, and biosensors.

Culture-based methods: These methods are based on the isolation and identification of *S. enterica* from the samples using selective and differential media, biochemical tests, and serological tests [38, 50]. The advantages of these methods are that they are simple, inexpensive, and widely available and that they can detect and identify viable *S. enterica* at the species and serotype level. They can also provide information on the antimicrobial susceptibility and resistance of the bacteria, which can be useful for therapy and control. The disadvantages of these methods are that they are time-consuming, labor-intensive, and prone to false negatives and false positives. They also require prior enrichment and culture of the samples, which can introduce contamination and variability [38, 50].

Molecular methods: These methods are based on the detection and identification of *S. enterica* from the samples using nucleic acid amplification techniques, such as polymerase chain reaction (PCR), real-time PCR, loop-mediated isothermal amplification (LAMP), or nucleic acid sequence-based amplification (NASBA) [38, 73]. The advantages of these methods are that they are rapid, sensitive, and specific and that they can detect and identify *S. enterica* at the species and serotype level, as well as the presence of virulence and resistance genes. They can also provide quantitative and real-time information about the bacteria, which can be useful for monitoring and diagnosis. The disadvantages of these methods are that they are complex, expensive, and require specialized equipment and trained personnel. They may also be affected by the quality and quantity of the DNA, the presence of inhibitors, and the specificity of the primers and probes [38, 73].

Immunological methods: These methods are based on the detection and identification of *S. enterica* from the samples using antigen-antibody reactions, such as enzyme-linked immunosorbent assay (ELISA), immunochromatographic assay (ICA), or immunofluorescence assay (IFA) [38, 52]. The advantages of these methods are that they are rapid, sensitive, and specific and that they can detect and identify *S. enterica* at the species and serotype level, as well as the presence of antigens and antibodies. They can also provide quantitative and semi-quantitative information about the bacteria, which can be useful for monitoring and diagnosis. The disadvantages of these methods are that they are complex, expensive, and require specialized equipment and trained personnel. They may also be affected by the stability and specificity of the antibodies, the interference and cross-reactivity of the antigens, and the calibration and validation of the assays [38, 52].

Phage-based methods: These methods are based on the detection and identification of *S. enterica* from the samples using bacteriophages, which are viruses that infect and lyse specific bacteria [38, 50]. The advantages of these methods are that they are rapid, sensitive, and specific and that they can detect and identify viable *S. enterica* directly from clinical samples without the need for culture or enrichment. They can also provide information on the susceptibility and resistance of the bacteria to phages, which can be useful for therapy and control. The disadvantages of these methods are that they are complex, expensive, and require specialized equipment and trained personnel. They may also be affected by the availability and specificity of the phages, the presence of other bacteria or viruses, and the environmental conditions [38, 50]. One of the most common phage-based methods is phage typing, which uses specific phages to identify *S. enterica* serotypes from the samples [74]. A recent study by [75] used phage typing to identify *S. enterica* serotypes in cattle fecal samples from dairy farms in Brazil and found that the most prevalent serotypes were Typhimurium, Enteritidis, and Dublin.

Biosensors: These devices are based on the detection and identification of *S. enterica* from the samples using biological recognition elements, such as antibodies, enzymes,

Method or technique	Advantages	Disadvantages
Culture methods	• Simple, inexpensive, and widely available • Provide viable and pure cultures of the bacteria • Allow further identification and characterization by biochemical and molecular methods	• Time-consuming, labor-intensive, and may not detect all the serotypes or distinguish closely related ones • Require viable and pure bacterial cultures, which may not always be available or feasible
Biochemical tests	• Simple, inexpensive, and widely available • Provide phenotypic and metabolic information of the bacteria • Useful for classification and diagnosis	• Time-consuming, labor-intensive, and may not identify and differentiate all the serotypes or distinguish closely related ones • Require pure and viable bacterial cultures, which may not always be available or feasible
Molecular methods	• Rapid, sensitive, and specific – detect and identify *Salmonella enterica* directly from clinical samples, without the need for culture or enrichment • Provide genotypic and phylogenetic information of the bacteria • Useful for typing and tracing	• Complex, expensive, and require specialized equipment and trained personnel • Affected by the quality and quantity of the DNA or RNA, the presence of inhibitors or contaminants, and the availability and accuracy of the primers, probes, or databases
Phage-based methods	• Rapid, sensitive, and specific • Detect and identify viable *S. enterica* directly from clinical samples without the need for culture or enrichment • Provide information on the susceptibility and resistance of the bacteria to phages • Useful for therapy and control	• Complex, expensive, and require specialized equipment and trained personnel • Affected by the availability and specificity of the phages, the presence of other bacteria or viruses, and the environmental conditions
Aptamer-based biosensors	• Rapid, sensitive, and specific – detect and identify *S. enterica* directly from clinical samples without the need for culture or enrichment • Provide quantitative and real-time information of the bacteria • Useful for monitoring and diagnosis	• Complex, expensive, and require specialized equipment and trained personnel • Affected by the stability and specificity of the biological recognition elements, the interference and noise of the physical or chemical transducers, and the calibration and validation of the devices

Table 6.
Advantages and disadvantages of different methods and techniques.

or DNA, coupled with physical or chemical transducers, such as optical, electrochemical, or piezoelectric sensors [38, 73]. The advantages of these devices are that they are rapid, sensitive, and specific and that they can detect and identify *S. enterica* directly from clinical samples without the need for culture or enrichment. They can also provide quantitative and real-time information on the bacteria, which can be useful for monitoring and diagnosis [72]. The disadvantages of these devices are that they are complex, expensive, and require specialized equipment and trained personnel. They may also be affected by the stability and specificity of the biological recognition elements, the interference and noise of the physical or chemical transducers, and the calibration and validation of the devices [38, 73]. One of the most novel biosensors is aptamer-based biosensors, which use synthetic nucleic acid molecules that bind to *S. enterica* with high affinity and specificity [76]. A recent study by [76, 77] used aptamer-based biosensors to detect *S. enterica* in milk samples and found that they had high sensitivity and specificity, as well as low detection limit and response time (**Table 6**).

4. Challenges and limitations of current methods and techniques

Some of the challenges and limitations of current methods and techniques for *S. enterica* detection are culture methods: These methods are time-consuming, labor-intensive, and may not be able to detect all the serotypes or distinguish closely related ones [44, 74]. Moreover, they require viable and pure bacterial cultures, which may not always be available or feasible. Biochemical tests are time-consuming, labor-intensive, and may not be able to identify and differentiate all the serotypes or distinguish closely related ones. Moreover, they require pure and viable bacterial cultures, which may not always be available or feasible. Molecular methods: These methods are complex, expensive, and require specialized equipment and trained personnel [76]. They may also be affected by the quality and quantity of the DNA or RNA, the presence of inhibitors or contaminants, and the availability and accuracy of the primers, probes, or databases. Phage-based methods are complex, expensive, and require specialized equipment and trained personnel. They may also be affected by the availability and specificity of the phages, the presence of other bacteria or viruses, and the environmental conditions [49].

5. Future trends and perspectives for improving identification methods and techniques

The future trends and perspectives for improving identification methods and techniques for *S. enterica* are:

i. Development of novel and improved molecular methods that can detect and identify *S. enterica* directly from clinical samples, without the need for culture or enrichment, such as isothermal amplification, nanopore sequencing, and CRISPR-based methods.

ii. Integration of multiple methods and techniques that can provide complementary and comprehensive information on *S. enterica*, such as culture, biochemical, molecular, phage-based, and biosensor methods.

iii. Application of bioinformatics and machine learning tools that can analyze and interpret the data generated by different methods and techniques, such as genomic, proteomic, and metabolomic data, and provide rapid and accurate identification and typing of *S. enterica*.

iv. Implementation of standardized and validated protocols and guidelines that can ensure the quality and reliability of the methods and techniques used for *S. enterica* detection, identification, and characterization [55, 56, 78].

v. Evaluation of the cost-effectiveness and feasibility of the methods and techniques used for *S. enterica* detection, identification, and characterization in different settings and scenarios, such as laboratories, farms, food industries, and public health agencies.

6. Conclusion

This chapter reviews the current methods and techniques for *S. enterica* detection in cattle, such as culture, biochemical, molecular, phage-based, and biosensor methods, and compares their advantages and disadvantages. We have highlighted several recent and relevant methods and techniques that have various challenges and limitations, such as low sensitivity and specificity, high cost and complexity, and the need for viable and pure bacterial cultures. Therefore, we have discussed the future trends and perspectives of novel and improved methods and techniques that can overcome these challenges and provide reliable and accurate information on *S. enterica* serotypes in cattle. We have highlighted some recent and relevant methods and techniques, such as phage typing and aptamer-based biosensors, and evaluated their potential and feasibility for *S. enterica* detection in cattle. We have also suggested some directions and recommendations for further research and development in this field, such as the integration and optimization of different methods and techniques, the validation and standardization of the methods and techniques, and the application and dissemination of the methods and techniques. We hope that this chapter can serve as a useful reference and guide for scientists, healthcare professionals, and lay users who are interested in or involved in the detection and identification of *S. enterica* in cattle.

Author details

Abdul Kabir[1*], Momin Khan[2] and Anees Ur Rahman[1]

1 Department of Veterinary Microbiology Faculty of Animal Husbandry and Veterinary Sciences, Sindh Agriculture University, Tandojam, Pakistan

2 Livestock and Dairy Development Department, Khyber Pakhtunkhwa, Pakistan

*Address all correspondence to: kabirvet32@gmail.com

References

[1] Knodler LA, Elfenbein JR. *Salmonella enterica*. Trends in Microbiology. 2019;**27**(11):964-965

[2] Carruthers L. Trends in antimicrobial resistant non-typhoidal salmonella before and after the food and drug administration's guidance for industry# 213, United States 2013-2020 [doctoral dissertation] Yale University. 2022

[3] Majowicz SE, Musto J, Scallan E, Angulo FJ, Kirk M, O'Brien SJ, et al. The global burden of nontyphoidal Salmonella gastroenteritis. Clinical Infectious Diseases. 2010;**50**(6):882-889

[4] Harbottle H, White DG, McDermott PF, Walker RD, Zhao S. Comparison of multilocus sequence typing, pulsed-field gel electrophoresis, and antimicrobial susceptibility typing for characterization of *Salmonella enterica* serotype Newport isolates. Journal of Clinical Microbiology. 2006;**44**(7):2449-2457

[5] Adem J, Bushra E. Bovine Salmonellosis and its public health importance: A review. Advances in Life Science and Technology. 2016;**44**:62-71

[6] Yada EL. A review on: Salmonellosis and its economic and public health significance. International Journal of Microbiological Research. 2023;**14**(2):21-33

[7] Senbeta TA. Epidemiology and public health importance of Bovine Salmonellosis. Journal Healthcare Treatment Development (JHTD). 2023;**3**(04):11-21. ISSN: 2799-1148

[8] Holschbach CL, Peek SF. Salmonella in dairy cattle. Veterinary Clinics: Food Animal Practice. 2018;**34**(1):133-154

[9] Wray C. Mammalian salmonellosis. In: Handbook of Zoonoses. 2nd ed. Section A. Boca Raton, Florida, United States: CRC Press; 2019. pp. 289-302

[10] Andino A, Hanning I. *Salmonella enterica*: Survival, colonization, and virulence differences among serovars. The Scientific World Journal. 2015;**2015**. Article ID 520179

[11] Zhou L, Pollard AJ. A fast and highly sensitive blood culture PCR method for clinical detection of *Salmonella enterica* serovar Typhi. Annals of Clinical Microbiology and Antimicrobials. 2010;**9**:1-7

[12] Rostagno MH, Gailey JK, Hurd HS, Mckean JD, Leite RC. Culture methods differ on the isolation of *Salmonella enterica* serotypes from naturally contaminated swine fecal samples. Journal of Veterinary Diagnostic Investigation. 2005;**17**(1):80-83

[13] Harvey RB, Anderson RC, Farrington LA, Droleskey RE, Genovese KJ, Ziprin RL, et al. Comparison of GN Hajna and tetrathionate as initial enrichment for salmonellae recovery from swine lymph nodes and cecal contents collected at slaughter. Journal of Veterinary Diagnostic Investigation. 2001;**13**(3):258-260

[14] Jameson JE. A discussion of the dynamics of Salmonella enrichment. Epidemiology & Infection. 1962;**60**(2):193-207

[15] Barco L, Lettini AA, Ramon E, Longo A, Saccardin C, Pozza MCD, et al. A rapid and sensitive method to identify and differentiate *Salmonella enterica* serotype Typhimurium and *Salmonella enterica* serotype 4, [5], 12: I:

By combining traditional serotyping and multiplex polymerase chain reaction. Foodborne Pathogens and Disease. 2011;**8**(6):741-743

[16] Lin L, Zheng Q, Lin J, Yuk HG, Guo L. Immuno-and nucleic acid-based current technique for Salmonella detection in food. European Food Research and Technology. 2020;**246**:373-395

[17] Hu L, Deng X, Brown EW, Hammack TS, Ma LM, Zhang G. Evaluation of Roka Atlas Salmonella method for the detection of salmonella in egg products in comparison with culture method, real-time PCR and isothermal amplification assays. Food Control. 2018;**94**:123-131

[18] Wattiau P, Boland C, Bertrand S. Methodologies for *Salmonella enterica* subsp. enterica subtyping: Gold standards and alternatives. Applied and Environmental Microbiology. 2011;**77**(22):7877-7885

[19] Wu S, Hulme JP. Recent advances in the detection of antibiotic and multi-drug resistant salmonella: An update. International Journal of Molecular Sciences. 2021;**22**(7):3499

[20] Rusul G, Chuah LO. Molecular methods for the detection and characterization of food-borne pathogens. Advances in Food Biotechnology. 2015:471-498

[21] Gorski L, Parker CT, Liang A, Cooley MB, Jay-Russell MT, Gordus AG, et al. Prevalence, distribution, and diversity of *Salmonella enterica* in a major produce region of California. Applied and Environmental Microbiology. 2011;**77**(8):2734-2748

[22] Hanson DL, Loneragan GH, Brown TR, Edrington TS. Salmonella prevalence varies over time and space in three large, adjacent cattle operations in the Southwestern United States. Frontiers in Animal Science. 2022;**3**:878408

[23] Nataro JP, Bopp CA, Fields PI, Kaper JB, Strockbine NA. Escherichia, shigella, and salmonella. Manual of Clinical Microbiology. 2011;**2011**:603-626

[24] Mkangara M. Prevention and control of human *Salmonella enterica* infections: An implication in food safety. International Journal of Food Science. 2023;**2023**. Article ID 8899596

[25] D'Angelo L, Vecchio D, Cozza D, La Tela I, Carullo MR, Menozzi I, et al. Identification of a New Serovar of *Salmonella enterica* in Mediterranean Buffalo calves (Bubalus bubalis). Animals. 2022;**12**(2):161

[26] Gebeyehu A, Taye M, Abebe R. Isolation, molecular detection and antimicrobial susceptibility profile of Salmonella from raw cow milk collected from dairy farms and households in southern Ethiopia. BMC Microbiology. 2022;**22**(1):1-10

[27] Hong S, Rovira A, Davies P, Ahlstrom C, Muellner P, Rendahl A, et al. Serotypes and antimicrobial resistance in *Salmonella enterica* recovered from clinical samples from cattle and swine in Minnesota, 2006 to 2015. PLoS One. 2016;**11**(12):e0168016

[28] Fegan N, Vanderlinde P, Higgs G, Desmarchelier P. Quantification and prevalence of salmonella in beef cattle presenting at slaughter. Journal of Applied Microbiology. 2004;**97**(5):892-898

[29] The Centers for Disease Control and Prevention (CDC) is headquartered in Atlanta, Georgia, U.S; 2021

[30] Münster P, Pöppel L, Antakli A, Müller-Doblies D, Radko D, Kemper N.

The detection of Salmonella enteritidis on German layer farms after cleaning and disinfection. Animals. 2023;**13**(16):2588

[31] Clothier K, Anderson M. Evaluation of bovine abortion cases and tissue suitability for identification of infectious agents in California diagnostic laboratory cases from 2007 to 2012. Theriogenology. 2016;**85**(5):933-938

[32] Schilling AK, Mazzamuto MV, Romeo C. A review of non-invasive sampling in wildlife disease and health research: What's new? Animals. 2022;**12**(13):1719

[33] McDonough SP, Southard T, editors. Necropsy Guide for Dogs, Cats, and Small Mammals. Hoboken, New Jersey, United States: John Wiley & Sons; 2017

[34] Arndt SS, Mazlan NH, van t Klooster JG, van t Lith HA, Kirchhoff S, Hoekman MFN, et al. Comparison of 2 blood sampling methods in mice to increase animal welfare and the reliability of experimental results. In: 68th AALAS National Meeting. Austin, Texas, USA: The American Association for Laboratory Animal Science (AALAS); 2017. p. 72

[35] Neculai-Valeanu AS, Ariton AM. Udder health monitoring for prevention of bovine mastitis and improvement of milk quality. Bioengineering. 2022;**9**(11):608

[36] Weese JS, Blondeau J, Boothe D, Guardabassi LG, Gumleyg N, Papichh M, et al. International Society for Companion Animal Infectious Diseases (ISCAID) guidelines for the diagnosis and management of bacterial urinary tract infections in dogs and cats. Journal of Japanese Association of Veterinary Nephrology and Urology. 2021;**13**(1):46-63

[37] Rostagno MH, Hurd HS, McKean JD, Ziemer CJ, Gailey JK, Leite RC. Preslaughter holding environment in pork plants is highly contaminated with salmonella enterica. Applied and Environmental Microbiology. 2003;**69**(8):4489-4494

[38] Lutz JK, Crawford J, Hoet AE, Wilkins JR III, Lee J. Comparative performance of contact plates, electrostatic wipes, swabs and a novel sampling device for the detection of Staphylococcus aureus on environmental surfaces. Journal of Applied Microbiology. 2013;**115**(1):171-178

[39] International Society for Biological and Environmental Repositories (ISBER). Collection, storage, retrieval and distribution of biological materials for research. Cell Preservation Technology. 2008;**6**(1):3-58

[40] Eriksson E, Aspan A. Comparison of culture, ELISA and PCR techniques for salmonella detection in faecal samples for cattle, pig and poultry. BMC Veterinary Research. 2007;**3**:1-19

[41] Koyuncu S, Haggblom P. A comparative study of cultural methods for the detection of Salmonellain feed and feed ingredients. BMC Veterinary Research. 2009;**5**(1):1-10

[42] Daquigan N, Grim CJ, White JR, Hanes DE, Jarvis KG. Early recovery of salmonella from food using a 6-hour non-selective pre-enrichment and reformulation of tetrathionate broth. Frontiers in Microbiology. 2016;**7**:2103

[43] Part A, Part B. RVS Enrichment Broth (BL017H-10ML-25BL). red, 35, 70

[44] Karolenko CE, Bhusal A, Gautam D, Muriana PM. Selenite cystine agar for enumeration of inoculated salmonella serovars recovered from stressful

conditions during antimicrobial validation studies. Microorganisms. 2020;**8**(3):338

[45] Morincigo MA, Martinez-Manzanares E, Muncoz A, Cornax R, Romero P, Borrego JJ. Evaluation of different plating media used in the isolation of salmonellas from environmental samples. Journal of Applied Bacteriology. 1989;**66**(4):353-360

[46] Kashani B, Kamil SA, Beigh AB, Shah SA, Wani BM, Kawoosa MS, et al. Isolation, identification and molecular characterization of *Salmonella enterica* subsp. enterica serovar Gallinarum from poultry farms of Central Kashmir, North India. 2021

[47] Dusch H, Altwegg M. Comparison of Rambach agar, SM-ID medium, and Hektoen enteric agar for primary isolation of non-typhi salmonellae from stool samples. Journal of Clinical Microbiology. 1993;**31**(2):410-412

[48] Van der Zee H. Media for the isolation of salmonella. In: Progress in Industrial Microbiology. Vol. 37. Amsterdam: Elsevier; 2003. pp. 195-208

[49] Warburton DW, Bowen B, Konkle A, Crawford C, Durzi S, Foster R, et al. A comparison of six different plating media used in the isolation of salmonella. International Journal of Food Microbiology. 1994;**22**(4):277-289

[50] Zhuang L, Gong J, Shen Q, Yang J, Song C, Liu Q, et al. Advances in detection methods for viable salmonella spp.: Current applications and challenges. Analytical Sciences. 2023;**39**(10):1643-1660

[51] Lee KM, Runyon M, Herrman TJ, Phillips R, Hsieh J. Review of salmonella detection and identification methods: Aspects of rapid emergency response and food safety. Food Control. 2015;**47**:264-276

[52] Hoorfar J, Malorny B, Abdulmawjood A, Cook N, Wagner M, Fach P. Practical considerations in design of internal amplification controls for diagnostic PCR assays. Journal of Clinical Microbiology. 2004;**42**(5):1863-1868

[53] Malorny B, Hoorfar J, Bunge C, Helmuth R. Multicenter validation of the analytical accuracy of salmonella PCR: Towards an international standard. Applied and Environmental Microbiology. 2003;**69**(1):290-296

[54] Van Belkum A, Tassios PT, Dijkshoorn L, Haeggman S, Cookson B, Fry NK, et al. Guidelines for the validation and application of typing methods for use in bacterial epidemiology. Clinical Microbiology and Infection. 2007;**13**:1-46

[55] Awang MS, Bustami Y, Hamzah HH, Zambry NS, Najib MA, Khalid MF, et al. Advancement in salmonella detection methods: From conventional to electrochemical-based sensing detection. Biosensors. 2021;**11**(9):346

[56] Shen Y, Xu L, Li Y. Biosensors for rapid detection of salmonella in food: A review. Comprehensive Reviews in Food Science and Food Safety. 2021;**20**(1):149-197

[57] Hadi J, Rapp D, Dhawan S, Gupta SK, Gupta TB, Brightwell G. Molecular detection and characterization of foodborne bacteria: Recent progresses and remaining challenges. Comprehensive Reviews in Food Science and Food Safety. 2023;**23**(2)

[58] Zhang X, Shi Y, Chen G, Wu D, Wu Y, Li G. CRISPR/Cas systems-inspired nano/biosensors for detecting infectious viruses and pathogenic bacteria. Small Methods. 2022;**6**(10):2200794

[59] Chen J, Karanth S, Pradhan AK. Quantitative microbial risk assessment

for Salmonella: Inclusion of whole genome sequencing and genomic epidemiological studies, and advances in the bioinformatics pipeline. Journal of Agriculture and Food Research. 2020;**2**:100045

[60] Kudirkiene E, Andoh LA, Ahmed S, Herrero-Fresno A, Dalsgaard A, Obiri-Danso K, et al. The use of a combined bioinformatics approach to locate antibiotic resistance genes on plasmids from whole genome sequences of *Salmonella enterica* serovars from humans in Ghana. Frontiers in Microbiology. 2018;**9**:1010

[61] Achtman M, Zhou Z, Alikhan NF, et al. Genomic diversity of *Salmonella enterica*-the UoWUCC 10K genomes project. Wellcome Open Research. 2020;**5**:223

[62] Sedeik ME, El-Shall NA, Awad AM, Elfeky SM, Abd El-Hack ME, Hussein EO, et al. Isolation, conventional and molecular characterization of Salmonella spp. from newly hatched broiler chicks. AMB Express. 2019;**9**:1-6

[63] Wang Y, Wang Y, Luo L, Liu D, Luo X, Xu Y, et al. Rapid and sensitive detection of Shigella spp. and Salmonella spp. by multiple endonuclease restriction real-time loop-mediated isothermal amplification technique. Frontiers in Microbiology. 2015;**6**:1400

[64] Mei X, Zhai X, Lei C, Ye X, Kang Z, Wu X, et al. Development and application of a visual loop-mediated isothermal amplification combined with lateral flow dipstick (LAMP-LFD) method for rapid detection of Salmonella strains in food samples. Food Control. 2019;**104**:9-19

[65] Maciorowski KG, Herrera P, Jones FT, Pillai SD, Ricke SC. Cultural and immunological detection methods for Salmonella spp. in animal feeds–A review. Veterinary Research Communications. 2006;**30**:127-137

[66] Kado CI, Heskett MG. Selective media for isolation of agrobacterium, corynebacterium, erwinia, pseudomonas and xanthomonas. Phytopathology. 1970;**60**(6):969-976

[67] Rigby J, Elmerhebi E, Diness Y, Mkwanda C, Tonthola K, Galloway H, et al. Optimized methods for detecting Salmonella Typhi in the environment using validated field sampling, culture and confirmatory molecular approaches. Journal of Applied Microbiology. 2022;**132**(2):1503-1517

[68] Goodman LB, McDonough PL, Anderson RR, Franklin-Guild RJ, Ryan JR, Perkins GA, et al. Detection of Salmonella spp. in veterinary samples by combining selective enrichment and real-time PCR. Journal of Veterinary Diagnostic Investigation. 2017;**29**(6):844-851

[69] Charlton BR, Walker RL, Kinde H, Bauer CR, Channing-Santiago SE, Farver TB. Comparison of a Salmonella enteritidis-specific polymerase chain reaction assay to delayed secondary enrichment culture for the detection of Salmonella enteritidis in environmental drag swab samples. Avian Diseases. 2005;**49**(3):418-422

[70] Atlas RM, Synder JW. Reagents, stains, and media: Bacteriology. Manual of Clinical Microbiology. 2011;**1**:272-303

[71] Ges NH. The detection of specified micro-organisms. In: Handbook of Microbiological Quality Control in Pharmaceuticals and Medical Devices. Vol. 86. London, United Kingdom: CRC Press; 2000

[72] Soria MA, Bueno DJ. Culture Based Methods to Detect Salmonella from

Different Poultry Products. New York: Nova Science Publishers; 2016. pp. 57-86

[73] Grimont PA, Weill FX. Antigenic formulae of the Salmonella serovars. WHO Collaborating Centre for Reference and Research on Salmonella. 2007;**9**:1-166

[74] Kosznik-Kwaśnicka K, Ciemińska K, Grabski M, Grabowski Ł, Górniak M, Jurczak-Kurek A, et al. Characteristics of a series of three bacteriophages infecting *Salmonella enterica* strains. International Journal of Molecular Sciences. 2020;**21**(17):6152

[75] Silva GB, Campos FV, Guimarães MC, Oliveira JP. Recent developments in lateral flow assays for Salmonella detection in food products: A review. Pathogens. 2023;**12**(12):1441

[76] Shin WR, Sekhon SS, Kim SG, Rhee SJ, Yang GN, Won K, et al. Aptamer-based pathogen monitoring for *Salmonella enterica* ser. Typhimurium. Journal of Biomedical Nanotechnology. 2018;**14**(11):1992-2002

[77] Bruce-Tagoe TA, Bhaskar S, Kavle RR, Jeevanandam J, Acquah C, Ohemeng-Boahen G, et al. Advances in aptamer-based biosensors for monitoring foodborne pathogens. Journal of Food Science and Technology. 2023;**60**:1-20

[78] Portmann AC, Fournier C, Gimonet J, Ngom-Bru C, Barretto C, Baert L. A validation approach of an end-to-end whole genome sequencing workflow for source tracking of listeria monocytogenes and *Salmonella enterica*. Frontiers in Microbiology. 2018;**9**:446

Section 2

Advanced Techniques for *Salmonella* Detection and Biocontrol

Chapter 4

Salmonella Infections: An Update, Detection and Control Strategies

Nirmal Kirti, Singha S. Krishna and Das Shukla

Abstract

Salmonella belonging to the *Enterobacteriaceae* family is commonly divided into Typhoidal and non-typhoidal group. Clinical manifestations can range from gastroenteritis, bacteremia or septicemia without gastrointestinal (GI) upset, enteric fever, chronic carriage to focal infections like osteomyelitis, meningitis, endocarditis etc. Conventional diagnostic tests like blood, stool, food sample culture have a long turnaround time. Chromogenic media modification reduces identification time. Apart from Widal test and ELISA, rapid point of care serological tests like Tubex TF, Typhidot are advantageous in basic set up but limited by modest accuracy in high burden settings. Apart from immunological assays, other diagnostic modalities include PCR, mass spectrophotometry, spectroscopy, optical phenotyping, biosensors etc. Rising drug resistance of *Salmonella* to first line Ampicillin, Chloramphenicol and Cotrimoxazole with increasing nonsusceptibility to fluoroquinolones and 3rd generation cephalosporin for past few decades is a public threat. Prevention and control measures include basic sanitation, safe water access, safe food handling, public education, physical, chemical, biocontrol methods, vaccination etc. Targeting the menace of antimicrobial resistance in *Salmonella* species needs a collaborative effort like One Health approach which optimizes the public health, animal and environmental health and reduce the dependency on antibiotics.

Keywords: *Salmonella*, typhoid, paratyphoid, non-typhoidal, diagnosis, drug resistance: prevention: control

1. Introduction

The bacterium *Salmonella* causes foodborne diseases. It is a gram-negative rod-shaped facultative anaerobic bacterium, predominantly motile with over 2600 serovars [1]. Foodborne diseases is a major public health problem, especially for young children. It is a ubiquitous microorganism which can easily survive in hardy environments. Salmonellosis is one of the major diarrheal diseases worldwide [1]. Apart from diarrhoeal diseases, Typhoidal strains can cause life-threatening enteric fever. In addition to the lack of sanitation and potable water, rising antimicrobial resistance is compounding to the problem [2]. In 1880, Karl Joseph Eberth isolated bacilli presumed as an agent of Typhoid fever from the spleen and mesenteric lymph nodes of patients. It was George Theodor Gaffky, an army surgeon and assistant to Robert Koch who managed to isolate the bacilli in pure culture. The microorganism was named *Eberthella Typhi* or *Eberth-Gaffky* bacillus. In 1885, a medical research scientist Theobald Smith in the department headed by a veterinary pathologist, Daniel Elmer Salmon, isolated a microbe called Hog-Cholera Bacillus from porcine intestines affected with swine fever. Joseph Leon Ligneres in the 1900s proposed the name *Salmonella* for the genus in honor of Daniel E. Salmon [3, 4].

2. Taxonomy and nomenclature

Salmonella is a complex group of bacteria. It belongs to Kingdom *Eubacteria*, phylum *Proteobacteria*, class *Gamma proteobacteria*, order *Enterobacterales*, family *Enterobacteriaceae* and Genus *Salmonella* [5]. The nomenclature of *Salmonella* is complex, controversial and ever-evolving. It amalgamates genetic, biochemical, and serological classification. Presently, the Centre for Disease Control and Prevention (CDC) uses nomenclature based on recommendations by the World Health Organization (WHO) Collaborating Centre [6]. Based on 16sRNA sequence analysis, this system divides *Salmonella* into two broad species *S. enterica* and *S. bongori*.

S. enterica is again categorized into six subspecies: *enterica* (I); *salamae* (II); *arizonae* (IIIa); *diarizonae* (IIIb); *houtenae* (IV); and *indica* (VI). Roman numerals (I to VI) have been designated to the six subspecies. Subspecies V does not appear in this classification as it was used to describe *S.enterica* subsp. *bongori* which has since been promoted to separate *Salmonella* species [6, 7]. **Figure 1** summarizes the taxonomical classification of *Salmonella.*

Apart from this, further classification of *Salmonella* subspecies is based on antigenic classification based on the somatic (O) or capsular (K) and flagellar (H) antigen detected by agglutination which was given by Kauffmann-White, now known as Kauffmann-White- Le Minor scheme [7]. Historically, serogroups based on O antigen were designated by alphabets e.g. A, B, C and so on. With the increasing number of serogroups identified, the alphabets were replaced with numbers. In certain Typhoidal serovars, Vi antigen is present superficially over the O antigen layer. An updated scheme with newer serotypes identified by the Word Health Organization Collaborating Centre for Reference and Research on *Salmonella* at the Pasteur

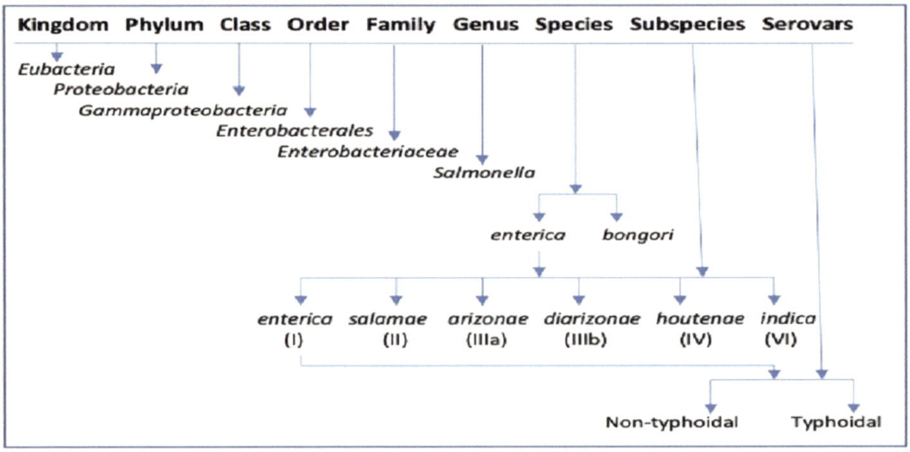

Figure 1.
Salmonella *taxonomical classification.*

Institute, Paris, France is reported in Research in Microbiology by Popoff et al. yearly, which has reported over 2600 serotypes or serovars [8]. The terminologies serotype and serovar seem to be used interchangeably. The Association of Public Health Laboratories and the World Health Organization prefer to use the term serovar in their documentation [9, 10]. The Rules of the Bacteriological Code also prefers serovar to serotype [10]. The naming of *Salmonella* strains before 1966 was based on host specificity, geographical origin or clinical features. All older serovars with specific names have been retained to this day with an underlying antigenic formula. Examples of some of the classical serovar names still used to these days are Agama (isolated from the Agama lizard feces), Agona (from cattle in the town Agonas of Ghana in 1952), Choleraesuis ("Swine cholera," Hog cholera cause), Dublin (isolated from the stool of a patient in Dublin, Ireland in 1929), Heidelberg (isolated from a patient in Heidelberg, Germany in 1933), Infantis (isolated from an infant in Connecticut, USA in 1943), Typhimurium ("mouse typhi," typhoid fever cause in mice) etc. [3, 6, 7].

Antigenic formula usually begins with subspecies written in Roman letters followed by somatic O antigens, flagellar H antigens phase 1 and phase 2, followed by other results if present. A colon separates major antigens and comma separates components of one antigen. Various symbols appearing in the antigenic formula make it complicated for example, square brackets like [ ] means the O or H antigen in question may be present or absent, ( ) parentheses means antigen is weakly agglutinable, { } curly bracket means antigens are mutually exclusive, hyphen – means particular flagellar phase is absent and underlined O factor means the factor is coded by bacteriophage (lysogenic strain). For example, antigenic formulae for the *Salmonella* Infantis is: I 6,7,14:r:1,5. Interpretation for the Infantis formula is that it belongs to subspecies *enterica* I, positive for 6,7,14 O antigen, positive for r H1 antigen and positive for 1 & 5 H2 antigen. A serotype name is not italicized but begins with capital first letter to differentiate from species name when writing a name of a particular named strain. When citing the name first time, "serotype" term should follow the genus name or the word "serotype" can be abbreviated to "ser" e.g. *Salmonella* serotype Typhi or *Salmonella* ser Typhi. In subsequent citations in the text, the name is shortened with the genus name followed by the serotype name with capital first letter, e.g., *Salmonella* Typhi *or S.* Typhi [8, 9].

3. Disease transmission

Salmonellosis is mostly foodborne. Consumption of contaminated meat, dairy and eggs is the cause as the bacteria mostly resides in the gut of animals and birds, the major sources being pigs, cattle and poultry. Transmission through the consumption of contaminated produce like fruits, vegetables, dried seeds as well as cereals has also been documented [11]. Person-to-person spread occurs in enteric fever via ingestion of food and water contaminated with human feces as Typhoidal fever lacks animal reservoirs. Food handlers especially, asymptomatic carriers of the bacteria with poor hygiene, sanitation and lack of clean water contribute to the spread of the disease [12]. The history of Mary Mallon "Typhoid Mary" is synonymous with the spread of the disease. The Irish-born American cook was a healthy carrier who was found to be responsible for the disease in hundreds of people [13]. The infective dose of bacteria has been reported to be less in non-typhoidal disease to the tune of 10^3 whereas approximately 10^5 bacilli is the infective dose for enteric fever [14, 15]. Various factors exist which increase the susceptibility of the host to Salmonellosis.

Children less than 5 years of age, elderly and immunocompromised cases are more susceptible. Hosts with conditions like achlorhydria, gastric surgery, inflammatory bowel disease, hemoglobinopathies, and impaired immune conditions like diabetes mellitus, AIDS, leukemia and lymphomas are more likely to get infected with a lower infective dose [14–17].

4. Pathogenesis and virulence factors

Salmonella harbors a variety of virulence factors to be fully pathogenic. The establishment of salmonellosis depends on the host cell invasion ability of the particular strain, intracellular proliferation, presence of a complete lipopolysaccharide coat and its toxin secretion ability. The virulence determinants singly or in combination facilitate the entry, attachment, colonization and invasion by overriding host defense mechanisms [18]. Virulence traits include flagella, capsule, plasmids, adhesion systems like adhesins, invasins, fimbriae, hemagglutinins, exotoxins, endotoxins, biofilm formation and type 3 secretion systems (T3SS) encoded on the *Salmonella* pathogenicity islands (SPI) [18, 19]. The invasion mechanism by *Salmonella* is partially understood. Following ingestion of the bacteria, the first bottleneck which acts against *Salmonella* colonization is the host gastric acidity where the pH drops as low as 1.5. Higher pH due to the use of proton pump inhibitors, antacids or achlorhydria along with food particles shields the bacteria and increases the risk of transmission [20].

The bacilli colonize the ileum and colon by attaching to the epithelial cells (M cells) by fimbriae. They enter through its epithelial lining by interacting with apical microvilli. Bacteria-mediated endocytosis occurs when the bacilli trigger the formation of ruffles on the cell membrane of M cells leading to the engulfment of the bacteria in vesicles. These vacuoles containing *Salmonella* reach the submucosa where they are phagocytosed by macrophages. The bacilli become non-susceptible to the lysosomal enzymes of the macrophages due to the organism's PhoP/PhoQ regulatory system, following which they spread via lymphatics and blood to other organs of the body [14, 15]. SPIs are gene clusters present on several chromosomal areas which encode for various virulence traits and are plasmid or chromosomally located usually, associated with mobile genetic elements. Several SPIs have been identified but SPI 1–5 have been closely associated with *Salmonella* serovars especially *S. enterica,* pathogenic to humans [21]. Host cell invasion including secretion of effector protein responsible for host cell function, encoding for T3SS and macrophage apoptosis induction is mediated by SPI-1 whereas replication within macrophages is done by SPI-2 by maintaining *Salmonella*-containing vacuole. SPI-2 is also responsible for encoding a second set of T3SS responsible for systemic infection. SPI-3 is reported to be responsible for encoding proteins required for long-term survival of bacilli, SPI-4 and SPI-5 encodes for genes associated with toxin secretion, survival within macrophages and several other sets of T3SS effector proteins [21–23]. The T3SS encoded by SPIs allows the effector proteins to give rise to a cascade sequence of activation of signal transduction and rearrangement of the cell cytoskeleton leading to ruffling of epithelial cell membrane for pinocytosis [24]. SPIs harbor various genes like *spvC* concerned with the formation of *Salmonella* containing vacuole (SCV) following engulfment which can escape host defense mechanism, *sopE* associated with activation of the signaling pathway for tissue damage and inflammation, suppression of signaling pathway for bacterial multiplication inhibition etc. [25, 26]. SPIs also harbor genes for encoding proteins involved in toxin mediation, virulence enhancement etc. The SCV prevents fusion with lysosomes and allows intracellular multiplication within macrophages. They spread to mesenteric lymph nodes and later reach the reticuloendothelial system (RES) via systemic circulation [24]. The invasion of various organs like gall bladder, liver, bone etc. depends on the interplay of serovars ingested and the effectiveness of the host immune system. Mostly, Salmonellosis causes acute diarrhea as the majority of the extra-intestinal bacilli are killed immediately [16, 27]. Mucosal inflammation causes release of numerous proinflammatory cytokines like IL-1, IL-6, IL-8, TNFα, IFN-γ, IL-12, IL-18 and IL-15 etc. [28]. Acute inflammatory response elicited by the pro-inflammatory cytokines causes intestinal ulceration along with inflammatory symptoms like fever, abdominal pain, diarrhea etc. *Salmonella* also triggers the activation of intestinal mucosal adenylate cyclase related to local generation of prostaglandin and enterotoxins which increases cyclic AMP, resulting in intestinal secretion. *Salmonella* escapes from the basal epithelial cells into lamina propria spreading systemically resulting in enteric fever [29].

5. Clinical features

Infections caused by *Salmonella* can be broadly classified into Typhoidal and non-typhoidal salmonellosis. It can present as gastroenteritis, enteric fever, other extra intestinal manifestations and chronic carriage [30].

5.1 Enteric fever

Enteric fever is caused by *S.* Typhi and *S.* Paratyphi A, B and C. Being indistinguishable from each other clinically, Typhoid and Paratyphoid fever are together called Enteric fever. Humans are the only reservoir for enteric fever. The incubation period of Typhoid fever is usually 6–30 days and that of Paratyphoid fever is typically 1–10 days [31]. Disease prodrome includes headache, malaise, loss of appetite, abdominal pain, diarrhea or constipation along with low-grade fever. Slowly, the fever rises from low grade to continuous high grade with the onset of the second week which is known as the step ladder pattern of fever [14, 15]. Pea soup diarrhea is usually seen in patients especially, children but immunosuppressed patients can present with constipation too. Other signs and symptoms include nausea, vomiting, relative bradycardia, rose spot on the trunk (in 30% of cases), hepatosplenomegaly, and myalgia and in some cases, cough and sore throat can also be seen. The dreaded complication occurring in 3% of the cases include gastrointestinal bleeding and intestinal perforation of peyer's patches at the terminal ileum usually observed in the third week. Other extraintestinal complications include myocarditis, metastasis to bone, joints, urinary tract etc. Gastrointestinal complications can present as pancreatitis, hepatitis and cholecystitis etc. Neurological complications include encephalopathy, meningitis, cerebellar ataxia and neuropsychiatric symptoms etc. [32, 33].

5.2 Non-typhoidal gastroenteritis

Non-typhoidal *Salmonella* (NTS) are *Salmonella* strains other than *S.* Typhi and *S.* Paratyphi. Primary reservoirs of NTS are usually wild, domesticated animals as well as birds. *S.* Typhimurium is the most common cause of NTS food poisoning [34]. The incubation period of NTS food poisoning is usually 6–12 hours as compared to days to weeks of Typhoidal incubation time. It is an acute inflammatory gastroenteritis characterized by fever, abdominal pain, diarrhea, nausea, vomiting etc. and is usually self-limiting. This requires no treatment but is indicated in cases with immunosuppressed conditions, children and elderly age groups as it is likely to get complicated [15, 34].

5.3 Invasive non-typhoidal *Salmonella* (iNTS) infections

The characteristic of iNTS infections is bacteremia by NTS without any focus of infection or gastroenteritis [35]. The most common iNTS strains are Choleraesuis, Dublin, Enteritidis, Typhimurium, Bovismorbificans, Arizona, and Gallinarum etc. Risk factors for bacteremia with other extra-intestinal manifestations include malnutrition, haemolytic anemia (Sickle cell anemia, Thalassemia), T-cell immunosuppression (AIDS, Hodgkin Lymphoma, Steroid Use, Transplantation, Chemotherapy), Schistosomiasis, Chronic granulomatous disease, mutations or acquired auto-immunity compromising IL-12, IFNγ function etc. [36]. The virulence determinants of iNTS are reported to be SPI-1 & SPI-2 responsible for an invasion via T3SS expression, survival of bacteria in macrophages and systemic manifestations; *spv* (*Salmonella* plasmid virulence) gene in iNTS is involved in macrophage cytoskeleton disruption, suppression of MAP kinase and NF-κB signaling pathway for prolonged survival of bacteria and host iron homeostasis interference [25, 36, 37]. Manifestations of iNTS infections include most commonly bacteraemia characterized by high-grade fever without formation of rose spots, pneumonia, endocarditis, osteomyelitis, prosthetic joint infections, meningitis, visceral abscess etc. [36].

5.4 Carrier state

Carriers are important from the public health point of view as they are the major cause of outbreaks. Temporary carriers are the cases who continue to shed Typhoid bacteria up to 3 months after disease resolution. It is seen in approximately 10% of cases. In chronic carriers, approximately 2–5% of cases fail to fully clear the infection even after a year of resolution of the disease [38]. Chronic carriers act as a reservoir of infection in the community as they intermittently shed the bacteria in stool or urine. Long-term survival of bacteria in carriers is due to the breach of the epithelial barrier, evasion of the host immune system and the forming of a niche in the host body. Mechanical barrier due to the sphincter of Oddi along with the flushing inability of bacteriostatic bile by the host is associated with the long-term carriage of the pathogen [39]. In contrast to Typhoidal *Salmonella*, carriage is uncommon, approximately seen in only 0.1% of cases with NT Salmonellosis [34].

6. Epidemiology

World Health Organization reported Typhoid infections affecting approximately 9 million people with over 110,000 deaths as of 2019. Typhoid was mostly reported from the developing nations of Southeast Asia, Africa, the Eastern Mediterranean and the Western Pacific Region [2]. The European countries and USA reported a low number of cases approximately less than 10 per 100,000 population annually with mostly travel-related history to enteric fever endemic countries [40]. Low-incidence countries like Israel noted causative agents of enteric fever to be *S.* Paratyphi in double the numbers as compared to *S.* Typhi [41]. Indian subcontinent documented the highest number of cases compared to other regions. The incidence rate for India, Pakistan and Indonesia per 100, 000 was 340, 573 and 148 [42]. Earlier in 2000, African countries reported a low number of Typhoid cases as compared to South Asian countries but in recent years, there are reports of growing numbers of Typhoid fever cases. Poor diagnostic infrastructures and fewer epidemiological studies could be associated with less reporting [42–44]. A low incidence of 0.38 per 100, 000 of Typhoid Fever was documented in China with cases mostly clustering in children <3 years of age and elderly >60 years of age [45]. Rising incidences of *S.* Paratyphi infections as compared to *S.* Typhi infections from Asian countries show the need for detailed epidemiological studies.

Non-typhoidal *Salmonella* (NTS) affects more than 150 million cases and deaths of approximately 60,000 cases globally, thus becoming one of the leading bacterial causes of diarrhea [46]. The highest incidences were reported from Sub-Saharan African (227/100,000) and European countries (102/100,000). American (23/100,000), South East Asia (21/100,000), Middle East (0.8/100,000) and North Africa (0.8/100,000) documented low incidences of NTS infections [46, 47]. *S.* Enteritidis was the most common serotype reported globally for causing NTS infections. Second and third most common isolates affecting human were *S.* Typhimurium and *S.* Newport respectively [48].

7. Detection

Salmonella can be diagnosed by various techniques starting from conventional methods to newer non-culture methods like molecular-based, immunological-based, molecular-based, mass spectrometry, spectroscopy, optical phenotyping, biosensor methods etc. [49, 50].

7.1 Culture based techniques

Even though culture-based methods are time-consuming, it is still the gold standard in diagnosis. Stools and suspected food products are the most commonly tested clinical samples. Apart from stool, other clinical samples are blood, body fluids, pus, bone marrow, duodenal aspirate, urine, punch biopsy samples of rose spots etc. [51]. The two most commonly done diagnostic tests for enteric fever are culture and Widal test. Bacterial culture depends on the site, age group, antimicrobial intake and stage of illness. Though blood culture is done commonly, the sensitivity is lower and variable (40–80%) as compared to bone marrow culture with a higher sensitivity of more than 80%. Also, blood culture sensitivity is higher in the first week as compared to subsequent weeks of infection and also gets significantly affected with antibiotics intake unlike bone marrow culture [52]. Blood and sterile body fluids can be cultured in blood culture bottles which can be conventional blood culture or automated blood culture bottles (Bactec/BacTAlert) which are further subcultured onto MacConkey agar. Culture methodology is different for contaminated samples like stool or food samples. The fecal samples on day 0 are cultured onto selective media like Xylose Lysine Deoxycholate (XLD) along with an enrichment broth like Selenite F Broth and incubated at 37°C for 18–24 hours. The enrichment broth is further subcultured onto selective media on Day 1. Enrichment broths like Selenite cystine broth, Gram Negative broth, and Tetrathionate broth are required for the selective enrichment of *Salmonella* over other microflora [53]. FDA's Bacteriological Analytical Manual (BAM) recommends pre-enriching food samples in non-selective media like buffered peptone water followed by Rappaport Vassialidis soy (RVS) broth and Muller Kauffmann tetrathionate-novobiocin and finally plating onto two selective media [49]. Various selective media for *Salmonella* apart from XLD agar are Hecktoen Enteric, Deoxycholate Citrate, *Salmonella Shigella*, Brilliant Green, Bismuth Sulfite agar etc. [54]. The suspicious-looking *Salmonella* colonies are identified by automated or traditional biochemical methods. The traditional biochemical tests include sugar fermentation, urease, citrate, decarboxylase, indole test, H2S production, MRVP, and nitrate etc. Isolates with presumptive biochemicals are confirmed with serotyping with O and flagellar H antisera and additional biochemical tests may be required in inconclusive results [53, 55]. Automated identification methods which are commonly used include Vitek-2, Microscan Walkaway system, MALDI-TOF etc. Modifications to conventional media like fluorogenic and chromogenic media e.g. SM-ID, BBL CHROMagar *Salmonella* and Rambach agar etc. decrease the turnover time by 1 day due to direct detection and identification [49, 53].

7.2 Non-culture-based techniques

Immunological assays: The Widal test is a routinely used serological test in high-burden countries. It is an agglutination test quantifying antibodies against lipo-polysaccharide O and flagellar H antigen. Though very convenient for low resource settings for its simplicity, the pitfall of this test primarily lies in the need for cautious interpretation of the results, limited sensitivity, specificity and difficulty distinguishing current and past infections [54]. Single unpaired tests showed poor sensitivity and specificity, especially in the early course of infection. Paired serological testing improves diagnostic accuracy but is limited in its clinical utility [51, 54]. Unlike the conventional Widal agglutination test, newer techniques in the immunological assay are based on the principle of *Salmonella* antigen capture by monoclonal or polyclonal antibodies. Enzyme-linked immunosorbent assays (ELISA) have been used for detection as well as quantification of antibodies targeting O, H, Vi antigens [51]. The diagnostic accuracy of ELISA was found to be better than the Widal test in comparative studies and [56, 57]. However, in the early course of the infection, ELISA done with single unpaired sera suffers from the same limitations of specificity as that of the Widal test [51, 58]. Conventional ELISA in resource-limited settings poses the problem of costly laboratory equipment and the need for trained personnel [59]. Recent advances like manipulation with gold nanoparticles in ELISAs have improved the results by lowering the limit of detection [60]. One of the simplest detection methods is the latex agglutination test where visible agglutination will be seen when antigens coated on *Salmonella* will bind with antibodies coated on latex particles [49]. Even if outdated, it is still a simple and reliable test for confirmatory analysis of isolates [49, 61]. Commercially available rapid diagnostic kits like Tubex TF immunoassay and Typhidot immunochromatographic test (ICT) have been developed as point-of-care tests for enteric fever [62, 63]. Tubex TF is an inhibition magnetic binding immunoassay which is a semi-quantitative test based on visual interpretation of color development. The principle is based on the inhibition of the reaction of magnetic particles coated with antigens and antibody-coated colored latex particles by IgM antibody against *S.* Typhi O9 present in the serum. Tubex TF showed a sensitivity of 56–95% and a specificity of 72–95% [62]. Typhidot is a miniaturized dot enzyme immunoassay that detects IgG and IgM antibodies to *S.* Typhi outer membrane protein (OMP) antigen. Typhidot- M detects recent infection by detecting IgM antibody against *S.* Typhi after IgG inactivation [63]. It showed moderate sensitivity and specificity of 56–84% and 31–97% respectively [62]. None of them yielded satisfactory results when tested in high-burden countries [64]. Although cheap and simple, clinicians in low-resource endemic settings should cautiously rely on them as early diagnostic tools.

Molecular Assays: Molecular techniques broadly use the principle of DNA/RNA detection of specific pathogens by target or signal amplification or non-amplification methods [49]. PCR is an excellent tool for pathogen detection. The most commonly used PCR targets are flagellum genes like *fliC-d* for *S.* Typhi, *fliC-a* for *S.* Paratyphi A etc. Other targets are the invasion gene (*invA*), tetrathionate reductase gene (*ttr*), *Salmonella* enterotoxin gene (*stn*), outer membrane porin F gene (*ompF*), hyperinvasive locus A (*hilA*), virulence plasmid gene (*spvC*) etc. [65]. Isothermal amplification methods are also effective tools for detecting *Salmonella* like NASBA i.e. nucleic acid sequence-based amplification, RPA i.e. Recombinase polymerase amplification, LAMP i.e. Loop-mediated isothermal amplification etc. [49]. Other examples of molecular diagnostic modalities include DNA Microarray and WGS i.e. Whole Genome Sequencing. Whole genome sequencing has the upper hand over others in its ability to sequence the entire genome of the bacteria. Though having high specificity and sensitivity, high cost and requirement of well-trained personnel is a major drawback [49, 50].

Mass Spectrometry-Based and Spectroscopy Method: MALDI-TOF identifies bacteria after bioinformatics profiling from a library of spectral data from the mass spectral fingerprints which is generated from the isolate. Apart from MALDI-TOF, Liquid chromatography-mass spectrometry (LC-MS) works on a similar principle but is slower than MALDI-TOF [66]. Raman spectroscopy has been increasingly applied to pathogen diagnosis. A Raman spectrum is formed when scattered light from an analyte is analyzed. The Raman spectroscopy has been enhanced by combination with artificial intelligence and in conjunction with nanoparticles like Surface-Enhanced Raman Spectroscopy (SERS). Other rapid spectroscopy techniques include Near Infrared Spectroscopy (NIR), High Spectral Imaging (HSI), and Fournier Transformed Infrared Spectroscopy (FITR) etc. Cross-reactivity in identifying *Salmonella* serovars can be lessened in spectroscopy-based methods but some disadvantages are fluorescent background, signal saturation, cost and requirement of trained personnel [49, 50, 67, 68].

Optical phenotyping methods like BARDOT (Bacterial Rapid Detection using Optical Scattering Technology) and BISLD (Bacteria Identification System by Light Diffraction) are available. The principle of the light scattering phenomenon is used where scattering images of colonies are compared with reference image libraries [69]. SELA-BARDOT technology uses on plate screening technique with laser which generates individual scatter signatures of multiple pathogens grown on SEL agar and provides results in real time [70]. This optical phenotyping method is non-destructive, has less turnaround time and is economical due to no requirement for any probes or reagents [70]. BISLD (Bacteria Identification System by Light Diffraction) is an automated system for bacterial colony identification. Converging spherical waves on the colonies obtain Fresnel diffraction patterns along with 2-D transmission coefficient maps in order to identify multiple species of bacteria [71].

Biosensors: Biosensor is a compact analytical device which provides information on analytes with the help of a biological recognition element and a transducer [72]. The advantages of biosensors over conventional diagnostic techniques are their higher sensitivity, specificity, portability potential and rapid results [49, 72, 73]. The three main components of a biosensor are a bio receptor, a transducer and an electronic system. The role of the electronic system component in biosensor devices is signal amplification, processing data and displaying the output [49]. Bio receptors are the analytes which recognize the target and the commonly used bioreceptors in *Salmonella* studies are antibodies, aptamer, antimicrobial peptides, nucleic acid and bacteriophage with their own set of pros and cons [49, 73]. Aptamers are short single-stranded RNA or DNA oligonucleotides which bind the target molecules with high specificity and affinity [74]. Apart from their high specificity and affinity, aptamers have trumped other bio-receptors with various other advantages like their low price, high stability, low limit of detection, ease of production, customization etc. [73, 74]. The only disadvantage of aptamers is that it is nuclease sensitive [49, 74]. Antibody bioreceptors also have good affinity and specificity but it is expensive and unstable [49]. Bacteriophage bioreceptors have the potential to differentiate live from dead bacilli but their capture efficiency is low when dry and also, likely to lyse the bacterial cells in the process [73]. The nucleic acid bioreceptors have high stability but are only limited to genosensors [74]. Antimicrobial peptides demonstrated high affinity and sensitivity but the disadvantage of antimicrobial peptides lies in their poor specificity [49, 73]. The second important component of biosensors is transducers which convert the bioreceptor signals into measurable signals. One of the most common biosensors based on the transducer elements is electrochemical biosensors which can be further classified into amperometric, voltammetric, impedimetric and potentiometric biosensors [73, 75]. Optical biosensors have transducers which convert bioreceptor signals into optical ones and are classified as colourimetric, fluorescent, SERS, and Surface Plasmon Resonance (SPR) biosensors [73, 75, 76]. Other biosensors studied for *Salmonella* identification in food industries are Piezoelectric, Magnetic Relaxation Switching (MRS), Photothermal Biosensors etc. [49, 50, 72, 77–79]. Among all the established biosensors, electrochemical biosensors are the most popular and widely studied as they offer major advantages like high specificity, sensitivity, low cost per test, potential on-site testing, portability options etc. A few disadvantages include high initial investment for the instrument and bioreceptor-dependent sample preparation complexity [49, 73, 75]. The transduction principle of the amperometric biosensor is that it measures the current produced due to the redox reaction between two electrodes [80]. Voltammetric biosensors measure the current produced as a function of the varied potential applied in a controlled manner [75]. Impedimetric biosensors measure the electrical impedance generated by the analyte at the electrode surface [75]. The potentiometric transduction principle for biosensors is the measurement of electrical potential between a working electrode and a reference electrode under negligible or zero current flow [80]. Impedimetric biosensors among all the electrochemical biosensors are the most extensively studied due to their label-free quantitative analysis capacity of *Salmonella* [73]. Emerging onsite and user-friendly biosensors in the pipeline are nanomaterial biosensors, aptamer-based FRET biosensors, microfluidics-based, portable, smartphone-based biosensors and commercial biosensors like Biacore Q100 systems [49, 72, 73, 75–79]. The appeal of biosensors also lies in their utility in several fields like diagnosis, drug discovery, food safety check, environmental surveillance etc. [49, 73].

8. Antimicrobial treatment and drug resistance

Enteric fever was conventionally treated by first-line antibiotics consisting of ampicillin, chloramphenicol and cotrimoxazole. Strains resistant to these first-line antibiotics were termed as Multi-Drug Resistant (MDR) *Salmonella* which emerged in the Indian subcontinent, Africa and Southeast Asia in the 1980s causing numerous outbreaks with reports of emerging resistance as early as the 1960s [81, 82]. The shifting of the treatment to fluoroquinolones culminated in MDR strains showing increased resistance to them [83]. It led to a change in prescription to third-generation cephalosporin and azithromycin [84]. Soon, isolated reports of cephalosporin-resistant Typhoid emerged from Bangladesh [85]. The first outbreak of extensively drug-resistant (XDR) typhoid to all three first-line drugs along with fluoroquinolones and cephalosporin emerged from Pakistan in 2016, with reports arising from all endemic countries and even documented in USA [86, 87]. Azithromycin and Carbapenem remain effective against these XDR strains [87]. Though Azithromycin resistance is rare, selective antibiotic pressure eventually leads to the emergence of resistance. Azithromycin-resistant *S.* Typhi strains have been isolated from Bangladesh [88].

With vaccine coverage, *S.* Paratyphi infections are doubling in incidence as compared to that of *S.*Typhi. Though often clubbed together, the resistance profile of *S.* Paratyphi strains have been found to be different to that of S. Typhi. MDR trait is lower in *S.* Paratyphi than S. Typhi [89]. Surveillance for Enteric Fever in Asia Project reported MDR *S.* Paratyphi to be 2%, while none were reported from Nepal or Bangladesh [90]. Even if there is a low incidence of MDR strains, rising minimum inhibitory concentration to fluoroquinolones might point toward not using ciprofloxacin as the drug of choice [91, 92].

AMR is rising in NTS due to the heavy use of antibiotics in animal husbandry. Clinically NTS diarrhea does not need antibiotics unless invasive in nature, elderly, children or immunocompromised case. AMR rates jumped from 20% in the 1990s to 70% in high-income countries by the 2000s according to surveillance data [93]. Of the prevalent serotypes, *S.* Enteritidis was found to be more susceptible than *S.*Typhimurium. Ever since the discovery of *S.*Typhimurium DT104 resistant to ampicillin, chloramphenicol, streptomycin, sulfonamide, and tetracycline in the 1990s, such MDR phenotype is on the rise with up to 84% reported in surveillance data from USA [94]. Due to MDR *Salmonella* strains, cephalosporins and fluoroquinolones have become the drug of choice. NTS cephalosporin resistance has been reported to cause outbreaks with cefotaxime-resistant strains isolated from clinical samples in USA and European countries like Italy, Spain with similar trends emerging from Asian countries like Taiwan [93, 95, 96]. Upto 2.5% resistance to nalidixic acid and a 5% increase in MIC in fluoroquinolones have been reported in the USA [97]. Quinolone-resistant *S.*Typhimurium DT104 resulted in an outbreak due to its spread from animals to humans [98]. Alarming reports have emerged from Taiwan regarding 98% of a highly invasive NTS *S.* Cholerasuis found to be resistant to one or more antibiotics. *S.* Cholerasuis resistant to both ceftriaxone and fluoroquinolones have also been isolated [99, 100].

AMR in *Salmonella* spp. is either chromosomally mediated or plasmid-mediated. Some of the resistance mechanisms are antibiotic inactivation, alteration of drug targets or efflux pumps mediated by genes via point mutation or transfer by mobile genetic elements, phages or plasmids [101, 102]. Like other gram negative bacilli, TEM, SHV, CTX-M etc. are primary beta-lactamases (ESBL) which are enzymes that cleave beta lactam rings conferring resistance to penicillin and cephalosporin and the gebes transferred by plasmids like *IncC* and *IncI1* [103, 104]. Chloramphenicol resistance genes like *Cat1* and *Cat2* confer resistance by inactivating the drug by acetyltransferase enzyme [102]. Acquired mutation in quinolone determining region of topoisomerase genes, *gyrA, gyrB, parC* etc. in chromosomal DNA. Resistance mediated by plasmid is coded by qnr genes that protect the DNA gyrase, *aac(6')-ib-cr* that decreases binding and *qep*A, *oqx*AB genes that code for efflux systems [105, 106]. Tetracycline resistance is conferred by genes *tetA, tetB, tetG etc* which activates the efflux pump mechanism to drive the drug out of the cell [107]. The resistance genes of sulphonamides and trimethoprim are *sul1, sul2, sul3* and *dfrA1, dfrA12, dfrV, dhfr1, dhfrV, dhfrA7, dhfr12, dhfr13, dhfr17*etc.respectively [107–109]. *The A*zithromycin resistance mechanism reported in S.Paratyphi A was due to the R717 mutation in the *AcrB* protein [88].

9. Prevention and control

Prevention and control of *Salmonella* infections relies on vaccination, improving food and water sanitation, public education and adopting rational use of antibiotics. Vaccines for Typhoid fever have been available for a long time. Previously used whole-cell inactivated vaccines had numerous adverse reactions and are now no longer in use. One of the licensed vaccines is parenteral Vi capsular polysaccharide (Vi-CPS). Vi antibodies induce complement activation against *S.* Typhi and Vi-CPS is known to provide a protective antibody response against *S.* Typhi but there is no evidence of cross-protection against other serovars unless they express Vi. It can be only administered to individuals over 2 years of age [110, 111]. Conjugated Vi antigen with recombinant *Pseudomonas aeruginosa* Exotoxin A (Vi-rEPA) increases immunogenicity and can be administered in children below 2 years of age unlike Vi- CPS vaccine [110]. Oral live attenuated Ty21a vaccine utilizes *S. enterica* var. Typhi Ty21a strain which lacks UDP-galactant- 4-epimerase which self destructs after inducing an immune response. Studies show that it offers some cross-protection against *S.* Paratyphi B but none against *S.* Paratyphi A [110, 112]. Lack of substantial efficacy data especially below 2 years of age, waning protection post-vaccination, need for boosters and lack of vaccine schedule implementation are some of the factors behind low vaccine uptake in endemic countries [113]. Out of the several typhoid conjugate vaccines (TCV) licensed in India, two have been prequalified by WHO to be used in high-burden countries. TYPBAR TCV uses Ty21 polysaccharide with tetanus toxoid. It's shown to be 87% efficacious and approved for children and infants less than 2 years of age. The second licensed TCV is TYPHIBEV with similar immunogenicity to TYPBAR TCV. Vi polysaccharide obtained from *Citrobacter freundii* which is indistinguishable from *S.* Typhi is conjugated with diphtheria toxin CRM197 as carrier protein [110, 114].

Effective food biosecurity measures to prevent contamination by *Salmonella* spp. in animals and produces along with isolation and quarantine in animal farms are one of the strategies for prevention and control [115]. Quality control by Hazard analysis and critical control point (HACCP) ensures consumer safety concerning food products and the standards are reviewed every 5 years to ensure the relevance of standards and protocols [116]. National and regional epidemiological surveillance programs are an important tool for detection and rapid awareness to control disease outbreaks [117, 118]. Good sanitation practices involving access to clean water supply, proper food handling practices, hand hygiene practices etc. are keys to curbing *Salmonella* infections, especially in endemic developing countries [117].

Physical interventions for control of *Salmonella* with regards to food safety that have been evaluated alone or in combination with other methods include ultrasound, UV, high-pressure processing, irradiation, pulsed light, plasma, thermal processing, high-intensity pulsed electric fields, oscillating magnetic fields etc. [119–121]. Chemical decontamination methods include the use of biocides like chlorine-based chemicals, chlorine alternatives (e.g. organic acids, peracetic acid), gas treatments (e.g. ozone, chlorine dioxide) etc. [120–122]. A biocide is a formulation containing an active substance to destroy or control any harmful organism to human or animal health by means other than physical or mechanical action [123]. Biocides for disinfection purposes can be divided into oxidizing (e.g. chlorine, chlorine dioxide, iodine, peroxides, peracetic acid etc.) and non-oxidizing groups (e.g. quaternary ammonium compounds, amphoterics, aldehydes, phenolic compounds, biguanides, and acid anionic agents etc) [122, 124]. However, there are greater implications regarding the widespread use of antimicrobial biocide like the development of biocide resistance, dermal or respiratory exposure effects, environmental threat due to persistence in aquatic systems etc. [122].

One of the novel biocontrol measures includes the use of bacteriophages (phages). Bacteriophages are viruses that infect and replicate in bacterial cells to create new progeny by using the host machinery [125]. The potential of phages against the rise of antibiotic-resistant *Salmonella* is due to their bacteriolytic activity, stability at a wider range of salt concentrations, pH, temperature, environment-friendly, self-limiting nature, genetic amenability etc. [126]. Phage cocktails with narrow host ranges have been reported to be better at reducing bacterial counts in various foods as they target multiple strains as compared to the use of a single phage [127]. A few examples of phages used against *Salmonella* in the food industry are SalmonFREE^{R}, PhageGuardS^{TM}, SalmoFresh^{TM}, SalmoLyse^{R} etc. [117]. Several challenges that exist with phage therapy are the emergence of phage resistance, restricted host range, moderate bacteriolytic activity, lack of safety data and regulatory framework for widespread commercial use etc. [117, 128, 129]. To combat some of the challenges, Genetically engineered phages have been modified to produce depolymerase, quenching enzymes, and hydrolases to act against biofilms, capsules and cell walls to enhance their bacteriolytic power. Also, the species-restricted host range in the engineered phages enables them to target only the pathogenic microbe and protect the commensals [128]. Bacteriophage-derived endolysins and Virion-associated peptidoglycan hydrolases are enzymes which are produced at the end of lytic cycles which hydrolyse the peptidoglycan layer from the outside as well as within the pathogenic bacteria. They are believed to be one of the potential alternative therapies against multidrug-resistant bacteria in multiple fields like food safety, medicine etc. due to good efficiency, rapid action, specificity and low chances of developing resistance [130].

Phytobiotics are plant extracts like herbs and spices which prevent food spoilage from contaminants, increase food shelf life and improve fortified food quality. In animals, phytobiotics owing to their antibacterial effects and microbiota modulation are known to improve carcass quality, decrease pathogenic bacteria proliferation, stimulate the immune system etc. [131, 132]. Bioactive compounds in phytobiotics also alter bacterial cytoplasmic membranes and genetic machinery, disrupt iron uptake pathways etc. [133]. Some examples of herbs and spices used against *Salmonella* spp. include cinnamon, thyme, oregano, essential oil from *Salvia officinalis,* lemon, lemongrass etc. [117, 134].

Prebiotics, probiotics, synbiotics and postbiotics help prevent gut colonization by *Salmonella.* Prebiotics are currently defined as a substrate that is selectively utilized by host microbes conferring a health benefit [135]. The prebiotic foods promote good bacteria to form biofilms on intestinal epithelium which in turn prevent pathogenic bacteria adhesion [136]. Probiotics are living microorganisms that confer a health benefit to the host when administered in the right amount. Probiotics include species of *Lactobacillus*, *Bifidobacterium, Bacillus, Enterococcus* and *Pediococcus,* [137]. Probiotics confer health benefits by competing with pathogens, stimulating the host immune response and inhibiting pathogens by secretion of lactic acid etc. [137]. Synbiotics are a combination of probiotics and prebiotics. The synergistic effects of probiotics and prebiotics inhibit *Salmonella* or lessen its effect by modulating gut microbiota, decreasing pathogen count, inducing additive effects of the immune system etc. [138]. Another emerging term is the postbiotics. Postbiotic preparations consist of non-viable microbes, their cell fragments and metabolites which are beneficial to the health of the host [139]. Various bacterial metabolites that can be used in postbiotic preparations are enzymes, exopolysaccharides, plasmalogens, organic acids, short-chain fatty acids, peptides etc. [140]. Postbiotics in animals showed reduced *Salmonella* count as well as increased nutrient absorption [141].

Another alternative method that can target bacteria is nanoparticles. Nanoparticles are ultrafine colloidal particles ranging from 1 to 100 nm in diameter [142]. The antibacterial mechanisms of nanoparticles are still unclear but the current theories are induction of oxidative stress, metal ion release and non-oxidative pathways [142]. Various types of nanoparticles against the control of *Salmonella* include metal-based, carbon-based, lipid-based and polymeric nanoparticles. Polymeric nanoparticles like chitosan nanoparticles are bigger in size making them efficient for oral uptake. They can withstand stomach acid, increase antigen-presenting cell uptake, stimulate the immune system continuously and act as adjuvants [143]. Silver nanoparticles act as carriers of plant extracts, antibiotics, or vaccines which are delivered on attachment to the cell membrane of *Salmonella* leading to microbial dysfunction [144].

Tackling drug-resistant *Salmonella* requires a multi-pronged approach on several fronts. Antibiotic dependency needs to be curbed with alternative non-antibiotic approaches, especially in animal husbandry. Research needs to be done to find newer strategies and further studies are required for the already existing alternative control measures. Since rising drug resistance in *Salmonella*, affects not only public health but also animal and environmental health, interdisciplinary and international collaborative efforts need to be framed within the One Health approach [145]. One health strategy includes the improved understanding of antimicrobial resistance mechanism through continued education, training and communication; implementation of infection prevention measures, and improved sanitation; antimicrobial stewardship programs for rational use of antibiotics in humans and animals; increased investments for vaccines and non-antimicrobial-based interventions; use of multi-omics (metagenomics, transcriptomics, proteomics etc) as a surveillance tool to identify the origin of antimicrobial resistance genes, its pattern and epidemiology [145, 146].

10. Conclusion

The disease burden caused by *Salmonella* is substantial and it poses a threat to the public health. Due to its ubiquitous and hardy nature, it persists for weeks in the environment. The multi-drug resistant strains affecting humans having varied virulence factors is a concern on a global level. The emergence of extensively drug-resistant typhoidal *Salmonella* from southeast Asia as well as reports of resistant non-typhoidal *Salmonella* is challenging the effective treatment availability. AMR in zoonotically transferable *Salmonella* is a matter of grave concern because of the extensive use of antibiotics in the husbandry of livestock. One health approach with a view of integrating surveillance and collaboration across human, animal, environmental and other relevant sectors like food industries is one of the key measures in tackling the menace of AMR. Other prevention measures are improving sanitation concerning clean water, safe food handling practices etc. Vaccination for Typhoid has been present for 100 years but suffered from low uptake due to progressively waning efficacy post-vaccination. The pre-qualification of newly licensed TCVs by WHO has been encouraging as the vulnerable age groups of children can be vaccinated. Further research is required for developing vaccines against Paratyphi serovars and non-typhoidal *Salmonella*, for at-risk groups like HIV and other immunocompromised cases. There is a strong case for developing multivalent vaccines which can target *Salmonella* serovars simultaneously like trivalent vaccines for invasive NTS/typhoidal strains in Africa and bivalent enteric fever vaccines in Asia or quadrivalent vaccines to provide cross-protection against all major serotypes.

Alternatives to antibiotics are the need of the hour to combat rising antimicrobial resistance in *Salmonella*. A combination of alternative strategies like biocides, prebiotics, probiotics, bacteriophages etc. can reduce the dependency on antibiotics.

It is of vital importance to commit to research and development so that efficient and advanced methods can quickly detect antibiotic-resistant *Salmonella* outbreaks for quick response. Besides good prevention and control approaches, continued research in pathogen detection is also crucial for *Salmonella* mitigation. Conventional culture methods along with non-culture methods like immunological assay, PCR, mass spectrophotometry, spectroscopy, optical phenotyping etc. have their niche in the diagnostic armamentarium. A promising candidate in the *Salmonella* detection system is biosensors. Apart from medical diagnostics, the appeal of biosensors extends to other fields like environmental surveillance, ensuring food safety, drug discovery etc. Biosensors in the form of lab-on-chip devices which are rapid, highly sensitive, specific, reasonably priced, and miniaturized will be a highly welcome innovation. The commercialization of technology where biosensors are combined with artificial intelligence (AI) will be a scientific breakthrough in the future.

Conflict of interest

The authors declare no conflict of interest.

Author details

Nirmal Kirti*, Singha S. Krishna and Das Shukla
Department of Microbiology, University College of Medical Sciences and Guru Teg Bahadur Hospital, Delhi, India

*Address all correspondence to: doctorkirtinirmal@gmail.com

References

[1] World Health Organization (WHO). Salmonella (Non-Typhoidal) Factsheets. 2018. Available from: https://www.who.int/news-room/fact-sheets/detail/salmonella-(non-typhoidal) [Accessed: December 28, 2023]

[2] World Health Organization (WHO). Salmonella Typhoid Factsheets. 2023. Available from: https://www.who.int/news-room/fact-sheets/detail/typhoid#:~:text=Typhoid%20fever%20is%20a%20life,and%20spread%20into%20the%20bloodstream [Accessed: December 28, 2023]

[3] Ryan MP, O'Dwyer J, Adley CC. Evaluation of the complex nomenclature of the clinically and veterinary significant pathogen *Salmonella*. BioMed Research International. 2017;**2017**:3782182

[4] Hardy A. Food, hygiene, and the laboratory. A short history of food poisoning in Britain, circa 1850-1950. Social History of Medicine. 1999;**12**(2):293-311

[5] MacKenzie KD, Palmer MB, Köster WL, White AP. Examining the link between biofilm formation and the ability of pathogenic *Salmonella* strains to colonize multiple host species. Frontiers in Veterinary Science. 2017;**4**:138

[6] Popoff MY, Bockemühl J, Gheesling LL. Supplement 2001 (no. 45) to the Kauffmann–White scheme. Research in Microbiology. 2003;**154**(3):173-174

[7] Brenner FW, Villar RG, Angulo FJ, Tauxe R, Swaminathan B. *Salmonella* nomenclature. Journal of Clinical Microbiology. 2000;**38**:2465-2467

[8] Issenhuth-Jeanjean S, Roggentin P, Mikoleit M, Guibourdenche M, De Pinna E, Nair S, et al. Supplement 2008-2010 (no. 48) to the White–Kauffmann–Le Minor scheme. Research in Microbiology. 2014;**165**:526-530

[9] Grimont PA, Weill F-X. Antigenic formulae of the *Salmonella* serovars. 9th Ed. Paris, France: WHO Collaborating Centre for Reference and Research on *Salmonella*; 2007

[10] Trüper HG. Judicial Commission Of The International Committee On Systematics Of Prokaryotes. The type species of the genus Salmonella Lignieres 1900 is Salmonella enterica (ex Kauffmann and Edwards 1952) Le Minor and Popoff 1987, with the type strain LT2T, and conservation of the epithet enterica in Salmonella enterica over all earlier epithets that may be applied to this species. Opinion 80. International Journal of Systematic and Evolutionary Microbiol. Jan 2005;**55**(1):519-520

[11] Ehuwa O, Jaiswal AK, Jaiswal S. *Salmonella*, food safety and food handling practices. Food. 2021;**10**(5):907

[12] Kirchhelle C, Pollard AJ, Vanderslott S. Typhoid-from past to future. Clinical Infectious Diseases [Internet]. 2019;**69**(Suppl 5):S375-S376

[13] Marineli F, Tsoucalas G, Karamanou M, Androutsos G. Mary Mallon (1869-1938) and the history of typhoid fever. Annals of Gastroenterology. 2013;**26**(2):132-134

[14] Bronze MS, Greenfield RA, editors. Biodefence Principles and Pathogens. Greenfield Norfolk, United Kingdom: Horizon Bioscience; 2005

[15] Ryan KJ, Ray CG, editors. Sherris Medical Microbiology: An Introduction to Infectious Disease. 4th ed. New York: McGraw-Hill; 2004

[16] Giannella RA. Salmonella. In: Baron S, editor. Medical Microbiology. 4th ed. Galveston (TX): University of Texas Medical Branch at Galveston; 1996. p. 21

[17] Pucciarelli MG, García-Del Portillo F. Salmonella intracellular lifestyles and their impact on host-to-host transmission. Microbiology Spectrum. 2017;**5**(4):10.1128

[18] Sabbagh SC, Forest CG, Lepage C, Leclerc JM, Daigle F. So similar, yet so different: Uncovering distinctive features in the genomes of *Salmonella enterica* serovars Typhimurium and Typhi. FEMS Microbiology Letters. 2010;**305**(1):1-13

[19] Daigle F. Typhi genes expressed during infection or involved in pathogenesis. Journal of Infection in Developing Countries. 2008;**2**(6):431-437

[20] Smith JL. The role of gastric acid in preventing foodborne disease and how bacteria overcome acid conditions. Journal of Food Protection. 2003;**66**:1292-1303

[21] Marcus SL, Brumell JH, Pfeifer CG, Finlay BB. *Salmonella* pathogenicity islands: Big virulence in small packages. Microbes and Infection. 2000;**2**(2):145-156

[22] Amavisit P, Lightfoot D, Browning GF, Markham PF. Variation between pathogenic serovars within *Salmonella* pathogenicity islands. Journal of Bacteriology. 2003;**185**(12):3624-3635

[23] Van Asten AJA, Van Dijk JE. Distribution of "classic" virulence factors among *Salmonella* spp. FEMS Immunology and Medical Microbiology. 2005;**44**(3):251-259

[24] Monack DM, Mueller A, Falkow S. Persistent bacterial infections: The interface of the pathogen and the host immune system. Nature Reviews Microbiology. 2004;**2**:747-765

[25] Guiney DG, Fierer J. The role of the spv genes in *Salmonella* pathogenesis. Frontiers in Microbiology. 2011;**2**:129

[26] Röder J, Hensel M. Presence of SopE and mode of infection result in increased *Salmonella*-containing vacuole damage and cytosolic release during host cell infection by *Salmonella* enterica. Cellular Microbiology. 2020;**22**:e13155

[27] Eng S-K, Pusparajah P, AbMutalib N-S, Ser H-L, Chan K-G, Lee L-H. *Salmonella*: A review on pathogenesis, epidemiology and antibiotic resistance. Frontiers in Life Science. 2015;**8**(3):284-293

[28] Eckmann L, Kagnoff MF. Cytokines in host defense against *Salmonella*. Microbes and Infection. 2001;**3**(14-15):1191-1200

[29] Ahmad Bhat K, Manzoor T, Ahmad Dar M, Farooq A, Ahmad Allie K, Majeed Wani S, et al. Salmonella infection and pathogenesis [internet]. In: Enterobacteria. London, UK: IntechOpen; 2022

[30] Darby J, Sheorey H. Searching for *Salmonella*. Australian Family Physician. 2008;**37**(10):806-810

[31] Centre for Disease Control and Prevention. Typhoid and Paratyphoid Fever. 2023. Available from: https://www.cdc.gov/typhoid-fever/health-professional.html#:~:text=Confusion%2C%20delirium%2C%20and%20intestinal%20

perforation,10%20days%20for%20paratyphoid%20fever [Accessed: January 02, 2024]

[32] Parry CM, Hien TT, Dougan G, White NJ, Farrar JJ. Typhoid fever. The New England Journal of Medicine. 2002;**347**:1770-1782

[33] Kuvandik C, Karaoglan I, Namiduru M, Baydar. Predictive value of clinical and laboratory findings in the diagnosis of the enteric fever. The New Microbiologica. 2009;**32**:25-30

[34] Hohmann EL. Nontyphoidal salmonellosis. Clinical Infectious Disease. 2001;**15**(32):263-269

[35] Saphra I, Winter JW. Clinical manifestations of salmonellosis in man; an evaluation of 7779 human infections identified at the New York Salmonella Center. The New England Journal of Medicine. 1957;**256**(24):1128-1134

[36] Fierer J. Invasive non-typhoidal Salmonella (iNTS) infections. Clinical Infectious Diseases. 2022;**75**(4):732-738

[37] Agbor TA, McCormick BA. Salmonella effectors: Important players modulating host cell function during infection. Cellular Microbiology. 2011;**13**(12):1858-1869

[38] Gunn JS, Marshall JM, Baker S, Dongol S, Charles RC, Ryan ET. Salmonella chronic carriage: Epidemiology, diagnosis, and gallbladder persistence. Trends in Microbiology. 2014;**22**(11):648-655

[39] Gonzalez-Escobedo G et al. Chronic and acute infection of the gall bladder by *Salmonella* Typhi: Understanding the carrier state. Nature Reviews Microbiology. 2011;**9**:9-14

[40] Cooke FJ, Day M, Wain J, Ward LR, Threlfall EJ. Cases of typhoid fever imported into England, Scotland and Wales (2000-2003). Transactions of the Royal Society of Tropical Medicine and Hygiene. 2007;**101**:398-404

[41] Meltzer E, Yossepowitch O, Sadik C, Dan M, Schwartz E. Epidemiology and clinical aspects of enteric fever in Israel. The American Journal of Tropical Medicine and Hygiene. 2006;**74**:540-545

[42] Wain J, Hendriksen RS, Mikoleit ML, Keddy KH, Ochiai RL. Typhoid fever. Lancet. 2015;**385**(9973):1136-1145

[43] Crump JA. Progress in typhoid fever epidemiology. Clinical Infectious Diseases. 2019;**68**(Suppl 1):S4-S9

[44] Breiman RF, Cosmas L, Njuguna H, Audi A, Olack B, Ochieng JB, et al. Population-based incidence of typhoid fever in an urban informal settlement and a rural area in Kenya: Implications for typhoid vaccine use in Africa. PLoS One. 2012;**7**(1):e29119

[45] Gao XY, Tang QY, Liu FF, Song Y, Zhang ZJ, Chang ZR. Epidemiological characteristics of typhoid fever and paratyphoid fever in China, 2004-2020. Zhonghua Liu Xing Bing Xue Za Zhi. 2023;**44**(5):743-750

[46] Balasubramanian R, Im J, Lee JS, Jeon HJ, Mogeni OD, Kim JH, et al. The global burden and epidemiology of invasive non-typhoidal *Salmonella* infections. Human Vaccines & Immunotherapeutics. 2019;**15**(6):1421-1426

[47] Ao TT, Daugla DM, Toralta J, Ngadoua C, Fermon F, Page A-L, et al. Global burden of invasive nontyphoidal Salmonella disease. 2010(1). Emerging Infectious Diseases. 2015;**21**(6):941-949

[48] Galanis E, Lo Fo Wong DM, Patrick ME, Binsztein N, Cieslik A,

et al. Web-based surveillance and global Salmonella distribution, 2000-2002. Emerging Infectious Diseases. 2006;**12**:381-388

[49] Awang MS, Bustami Y, Hamzah HH, Zambry NS, Najib MA, Khalid MF, et al. Advancement in *Salmonella* detection methods: From conventional to electrochemical-based sensing detection. Biosensors (Basel). 2021;**11**(9):346

[50] Wang M, Zhang Y, Tian F, Liu X, Du S, Ren G. Overview of rapid detection methods for *Salmonella* in foods: Progress and challenges. Food. 2021;**10**(10):2402

[51] Andrews JR, Ryan ET. Diagnostics for invasive Salmonella infections: Current challenges and future directions. Vaccine. 2015;**33**:C8-C15

[52] Gilman RH, Terminel M, Levine MM, Hernandez-Mendoza P, Hornick RB. Relative efficacy of blood, urine, rectal swab, bone-marrow, and rose-spot cultures for recovery of Salmonella Typhi in typhoid fever. Lancet. 1975;**1**:1211-1213

[53] Carroll KC, Pfaller MA, Landry ML, McAdam AJ, Patel R, Richter SS, et al., editors. Manual of Clinical Microbiology. 12th ed. Washington, D.C.: ASM Press; 2019

[54] Dutta S, Sur D, Manna B, Sen B, Deb AK, et al. Evaluation of new-generation serologic tests for the diagnosis of typhoid fever: Data from a community-based surveillance in Calcutta, India. Diagnostic Microbiology and Infectious Disease. 2006;**56**:359-365

[55] Procop GW, Koneman EW. Koneman's Color Atlas and Textbook of Diagnostic Microbiology. 7th ed. Baltimore, MD: Wolters Kluwer; 2016

[56] Adhikari A, Rauniyar R, Raut PP, Manandhar KD, Gupta BP. Evaluation of sensitivity and specificity of ELISA against Widal test for typhoid diagnosis in endemic population of Kathmandu. BMC Infectious Diseases. 2015;**15**:523

[57] Fadeel MA, House BL, Wasfy MM, Klena JD, Habashy EE, Said MM, et al. Evaluation of a newly developed ELISA against Widal, TUBEX-TF and Typhidot for typhoid fever surveillance. Journal of Infection in Developing Countries. 2011;**5**(3):169-175

[58] House D, Chinh NT, Diep TS, Parry CM, Wain J, Dougan G, et al. Use of paired serum samples for serodiagnosis of typhoid fever. Journal of Clinical Microbiology. 2005;**43**(9):4889-4890

[59] Thiha A, Ibrahim F. A colorimetric enzyme-linked immunosorbent assay (ELISA) detection platform for a point-of-care dengue detection system on a lab-on-compact-disc. Sensors (Basel). 2015;**15**(5):11431-11441

[60] De La Rica R, Stevens MM. Plasmonic ELISA for the ultrasensitive detection of disease biomarkers with the naked eye. Nature Nanotechnology. 2012;**7**:821-824

[61] Thorns CJ, McLaren IM, Sojka MG. The use of latex particle agglutination to specifically detect Salmonella Enteritidis. International Journal of Food Microbiology. 1994;**21**:47-53

[62] Thriemer K, Ley B, Menten J, Jacobs J, van den Ende J. A systematic review and meta-analysis of the performance of two point of care typhoid fever tests, Tubex TF and Typhidot, in endemic countries. PLoS One. 2013;**8**:e81263

[63] Beig FK, Ahmad F, Ekram M, Shukla I. Typhidot M and Diazo test vis-à-vis blood culture and Widal test in the early diagnosis of typhoid fever in

children in a resource poor setting. The Brazilian Journal of Infectious Diseases. 2010;**14**(6):589-593

[64] Najib MA, Mustaffa KMF, Ong EBB, Selvam K, Khalid MF, Awang MS, et al. Performance of immunodiagnostic tests for typhoid fever: A systematic review and meta-analysis. Pathogens. 2021;**10**(9):1184

[65] Ricke SC, Kim SA, Shi Z, Park SH. Molecular-based identification and detection of Salmonella in food production systems: Current perspectives. Journal of Applied Microbiology. 2018;**125**(2):313-327

[66] Bell RL, Jarvis KG, Ottesen AR, McFarland MA, Brown EW. Recent and emerging innovations in Salmonella detection: A food and environmental perspective. Microbial Biotechnology. 2016;**9**:279-292

[67] Pandian S, Lakshmi SA, Priya A, Balasubramaniam B, Zaukuu J-LZ, Durgadevi R, et al. Spectroscopic methods for the detection of microbial pathogens and diagnostics of infectious diseases—An updated overview. PRO. 2023;**11**(4):1191

[68] Sun J, Xu X, Feng S, Zhang H, Xu L, Jiang H, et al. Rapid identification of Salmonella serovars by using Raman spectroscopy and machine learning algorithm. Talanta. 2023;**253**:123807

[69] Singh AK, Bettasso AM, Bae E, Rajwa B, Dundar MM, Forster MD, et al. Laser optical sensor, a label-free on-plate Salmonella enterica colony detection tool. MBio. 2014;**5**:e01019-e01013

[70] Abdelhaseib MU, Singh AK, Bhunia AK. Simultaneous detection of Salmonella enterica, Escherichia coli and Listeria monocytogenes in food using a light scattering sensor. Journal of Applied Microbiology. 2019;**126**(5):1496-1507

[71] Suchwalko A, Buzalewicz I, Wieliczko A, Podbielska H. Bacteria species identification by the statistical analysis of bacterial colonies Fresnel patterns. Optics Express. 2013;**21**(9):11322-11337

[72] Kirsch J, Siltanen C, Zhou Q, Revzin A, Simonian A. Biosensor technology: Recent advances in threat agent detection and medicine. Chemical Society Reviews. 2013;**42**(22):8733-8768

[73] Shen Y, Xu L, Li Y. Biosensors for rapid detection of Salmonella in food: A review. Comprehensive Reviews in Food Science and Food Safety. 2021;**20**(1):149-197

[74] Park KS. Nucleic acid aptamer-based methods for diagnosis of infections. Biosensors & Bioelectronics. 2017;**102**:179-188

[75] Silva NFD, Magalhães JMCS, Freire C, Delerue-Matos C. Electrochemical biosensors for Salmonella: State of the art and challenges in food safety assessment. Biosensors and Bioelectronics. 2018;**99**:667-682

[76] Khansili N, Rattu G, Krishna PM. Label-free optical biosensors for food and biological sensor applications. Sensors and Actuators B: Chemical. 2018;**265**:35-49

[77] Wang L, Wang R, Chen F, Jiang T, Wang H, Slavik M, et al. QCM-based aptamer selection and detection of Salmonella Typhimurium. Food Chemistry. 2017;**221**:776-782

[78] Yue X, Sun J, Yang T, Dong Q, Li T, Ding S, et al. Rapid detection of Salmonella in milk by a nuclear magnetic resonance biosensor based

on the streptavidin-biotin system and O-carboxymethyl chitosan target gadolinium probe. Journal of Dairy Science. 2021;**104**(11):11486-11498

[79] Lu Y, Shi Z, Liu Q. Smartphone-based biosensors for portable food evaluation. Current Opinion in Food Science. 2019;**28**:74-81

[80] Riu J, Giussani B. Electrochemical biosensors for the detection of pathogenic bacteria in food. TrAC Trends in Analytical Chemistry. 2020;**126**:115863

[81] Britto CD, Wong VK, Dougan G, Pollard AJ. A systematic review of antimicrobial resistance in *Salmonella* enterica serovar Typhi, the etiological agent of typhoid. PLoS Neglected Tropical Diseases. 2018;**12**:e0006779

[82] Levine MM, Simon R. The gathering storm: Is untreatable typhoid fever on the way? MBio. 2018;**9**(2):e00482-18

[83] Hirose K, Hashimoto A, Tamura K, et al. DNA sequence analysis of DNA gyrase and DNA topoisomerase IV quinolone resistance-determining regions of *Salmonella enterica* serovar Typhi and serovar Paratyphi A. Antimicrobial Agents and Chemotherapy. 2002;**46**:3249-3252

[84] Dahiya S, Sharma P, Kumari B, Pandey S, Malik R, Manral N, et al. Characterisation of antimicrobial resistance in *Salmonellae* during 2014-2015 from four centres across India: An ICMR antimicrobial resistance surveillance network report. Indian Journal of Medical Microbiology. 2017;**35**:61-68

[85] Saha SK, Talukder SY, Islam M, Saha S. A highly ceftriaxone-resistant *Salmonella* Typhi in Bangladesh. The Pediatric Infectious Disease Journal. 1999;**18**:387

[86] Ali Shah SA, Nadeem M, Syed SA, Fatima Abidi ST, Khan N, Bano N. Antimicrobial sensitivity pattern of Salmonella Typhi: Emergence of resistant strains. Cureus. 2020;**12**(11):e11778

[87] Chatham-Stephens K, Medalla F, Hughes M, et al. Emergence of extensively drug-resistant Salmonella Typhi infections among travelers to or from Pakistan-United States, 2016-2018. Morbidity and Mortality Weekly Report. 2019;**68**:11-13

[88] Hooda Y, Sajib MSI, Rahman H, et al. Molecular mechanism of azithromycin resistance among typhoidal *Salmonella* strains in Bangladesh identified through passive pediatric surveillance. PLoS Neglected Tropical Diseases. 2019;**13**:e0007868

[89] Rowe B, Ward LR, Threlfall EJ. Multidrug-resistant Salmonella Typhi: A worldwide epidemic. Clinical Infectious Diseases. 1997;**24**:S106-S109

[90] Qamar FN, Yousafzai MT, Dehraj IF, Shakoor S, Irfan S, Hotwani A, et al. Antimicrobial resistance in Typhoidal Salmonella: Surveillance for enteric fever in Asia project, 2016-2019. Clinical Infectious Diseases. 2020;**71**(Suppl 3):S276-S284

[91] Hasan R, Zafar A, Abbas Z, Mahraj V, Malik F, Zaidi A. Antibiotic resistance among Salmonella enterica serovars Typhi and Paratyphi A in Pakistan (2001-2006). Journal of Infection in Developing Countries. 2008;**2940**:289-294

[92] Saha S, Sajib MSI, Garrett D, Qamar FN. Antimicrobial resistance in Typhoidal Salmonella: Around the world in 3 days. Clinical Infectious Diseases. 2020;**71**(Suppl 2):S91-S95

[93] Lin-Hui S, Chiu C-H, Chu C, Ou JT. Antimicrobial resistance in nontyphoid *Salmonella* serotypes: A global challenge. Clinical Infectious Diseases. 2004;**39**(4):546-551

[94] Crump JA, Medalla FM, Joyce KW, Krueger AL, Hoekstra RM, Whichard JM, et al. Antimicrobial resistance among invasive nontyphoidal Salmonella enterica isolates in the United States: National antimicrobial resistance monitoring system, 1996 to 2007. Antimicrobial Agents and Chemotherapy. 2011;**55**:1148-1154

[95] Fey PD, Safranek TJ, Rupp ME, Dunne EF, Ribot E, Iwen PC, et al. Ceftriaxone-resistant Salmonella infection acquired by a child from cattle. The New England Journal of Medicine. 2000;**342**(17):1242-1249

[96] Threlfall EJ, Fisher IS, Berghold C, Gerner-Smidt P, Tschäpe H, Cormican M, et al. Antimicrobial drug resistance in isolates of Salmonella enterica from cases of salmonellosis in humans in Europe in 2000: Results of international multi-centre surveillance. Euro Surveillance. 2003;**8**(2):41-45

[97] Crump JA, Barrett TJ, Nelson JT, Angulo FJ. Reevaluating fluoroquinolone breakpoints for Salmonella enterica serotype Typhi and for non-Typhi Salmonellae. Clinical Infectious Diseases. 2003;**37**(1):75-81

[98] Mølbak K, Baggesen DL, Aarestrup FM, Ebbesen JM, Engberg J, Frydendahl K, et al. An outbreak of multidrug-resistant, quinolone-resistant salmonella enterica serotype typhimurium DT104. The New England Journal of Medicine. 1999;**341**(19):1420-1425

[99] Su LH, Chiu CH, Kuo AJ, Chia JH, Sun CF, Leu HS, et al. Secular trends in incidence and antimicrobial resistance among clinical isolates of Salmonella at a University Hospital in Taiwan, 1983-1999. Epidemiology and Infection. 2001;**127**(2):207-213

[100] Chiu CH, Su LH, Chu C, Chia JH, Wu TL, Lin TY, et al. Isolation of Salmonella enterica serotype choleraesuis resistant to ceftriaxone and ciprofloxacin. Lancet. 2004;**363**(9417):1285-1286

[101] Boerlin P, Reid-Smith RJ. Antimicrobial resistance: Its emergence and transmission. Animal Health Research Reviews. 2008;**9**:115-126

[102] Khan M, Shamim S. Understanding the mechanism of antimicrobial resistance and pathogenesis of *Salmonella enterica* serovar Typhi. Microorganisms. 2022;**10**(10):2006

[103] Riyaaz AAA, Perera V, Sivakumaran S, de Silva N. Typhoid fever due to extended spectrum β- lactamase-producing *Salmonella* enterica serovar Typhi: A case report and literature review. Case Reports in Infectious Diseases. 2018;**2018**:4610246

[104] Coipan CE, Westrell T, van Hoek AHAM, Alm E, Kotila S, Berbers B, et al. Genomic epidemiology of emerging ESBL-producing *Salmonella* Kentucky *bla*$_{\text{CTX-M-14b}}$ in Europe. Emerging Microbes & Infections. 2020;**9**(1):2124-2135

[105] Robicsek A, Jacoby GA, Hooper DC. The worldwide emergence of plasmid-mediated quinolone resistance. The Lancet Infectious Diseases. 2006;**6**(10):629-640

[106] Acheampong G, Owusu M, Owusu-Ofori A, et al. Chromosomal and plasmid-mediated fluoroquinolone resistance in human *Salmonella enterica* infection in Ghana. BMC Infectious Diseases. 2019;**19**:898

[107] Pavelquesi SLS, de Oliveira Ferreira ACA, Rodrigues ARM, de Souza Silva CM, Orsi DC, da Silva ICR. Presence of tetracycline and Sulfonamide resistance genes in *Salmonella* spp.: Literature review. Antibiotics (Basel). 2021;**10**(11):1314

[108] Thai TH, Hirai T, Lan NT, Shimada A, Ngoc PT, Yamaguchi R. Antimicrobial resistance of Salmonella serovars isolated from beef at retail markets in the north Vietnam. The Journal of Veterinary Medical Science. 2012;**74**(9):1163-1169

[109] Deekshit VK, Kumar BK, Rai P, Srikumar S, Karunasagar I, Karunasagar I. Detection of class 1 integrons in Salmonella Weltevreden and silent antibiotic resistance genes in some seafood-associated nontyphoidal isolates of Salmonella in south-west coast of India. Journal of Applied Microbiology. 2012;**112**(6):1113-1122

[110] Kumar S, Ghosh RS, Iyer H, Ray A, Vannice K, MacLennan C, et al. Typhoid in India: An age-old problem with an existing solution. The Journal of Infectious Diseases. 2021;**224**(Supplement_5):S469-S474

[111] Mastroeni P, Chabalgoity JA, Dunstan SJ, Maskell DJ, Dougan G. *Salmonella*: Immune responses and vaccines. Veterinary Journal. 2001;**161**:132-164

[112] Sears KT, Galen JE, Tennant SM. Advances in the development of Salmonella-based vaccine strategies for protection against Salmonellosis in humans. Journal of Applied Microbiology. 2021;**131**(6):2640-2658

[113] World Health Organization (WHO). Background Paper to SAGE on Typhoid Vaccine Policy Recommendations. Geneva: WHO; 2018

[114] Jin C, Gibani MM, Moore M, Juel HB, Jones E, Meiring J, et al. Efficacy and immunogenicity of a Vi-tetanus toxoid conjugate vaccine in the prevention of typhoid fever using a controlled human infection model of Salmonella Typhi: A randomised controlled, phase 2b trial. Lancet. 2017;**390**(10111):2472-2480

[115] Awad W, Ghareeb K. Some aspects of control of *Salmonella* infection in poultry for minimizing contamination in the food chain. World's Poultry Science Journal. 2014;**70**(3):519-530

[116] Awuchi CG. HACCP, quality, and food safety management in food and agricultural systems. Cogent Food and Agriculture. 2023;**9**(1):2176280

[117] Mkangara M. Prevention and control of human *Salmonella enterica* infections: An implication in food safety. International Journal of Food Science. 2023;**2023**:8899596

[118] Uelze L, Becker N, Borowiak M, et al. Toward an integrated genome-based surveillance of *Salmonella enterica* in Germany. Frontiers in Microbiology. 2021;**12**:article 626941

[119] Jones FT. A review of practical *Salmonella* control measures in animal feed. Journal of Applied Poultry Research. 2011;**20**:102-113

[120] Joint Food and Agriculture Organization (FAO)/World Health Organization (WHO) Expert Meetings on Microbiological Risk Assessment (JEMRA). Meeting Report. Prevention and Control of Microbiological Hazards in Fresh Fruits and Vegetables Part 4: Specific Commodities. 2023. Available from: https://www.who.int/publications/i/item/9789240077959 [Accessed: February 08, 2024]

[121] Agirdemir O, Yurdakul O, Keyvan E, Sen E. Effects of various chemical decontaminants on Salmonella Typhimurium survival in chicken carcasses. Food Science and Technology. 2021;**41**(2):335-342

[122] Bettencourt Cota J, Vieira-Pinto M, Oliveira M. Biocide use for the control of non-Typhoidal Salmonella in the food-producing animal scenario: A primary food production to fork perspective [internet]. In: Salmonella-Perspectives for Low-Cost Prevention, Control and Treatment. London, UK: IntechOpen; 2024

[123] European Parliament, European Council. Regulation (EU) No 528/2012 of the European Parliament and of the Council of 22 May 2012 Concerning the Making Available on the Market and Use of Biocidal Products Text with EEA Relevance. 2012

[124] Fisher J. Types of disinfectant. In: Encyclopaedia of Food Science. Food Technology and Nutrition. Cambridge, Massachusetts, USA: Academic Press; 1993. pp. 1382-1385

[125] Zhou WY, Sun SF, Zhang YS, et al. Isolation and characterization of a virulent bacteriophage for controlling *Salmonella* Enteritidis growth in ready-to-eat mixed-ingredient salads. Journal of Food Protection. 2021;**84**:1629-1639

[126] Albino LA, Rostagno MH, Hungaro HM, Mendonça RC. Isolation, characterization, and application of bacteriophages for *Salmonell*a spp. biocontrol in pigs. Foodborne Pathogens and Disease. 2014;**11**:602-609

[127] Islam MS, Zhou Y, Liang L, Nime I, Liu K, Yan T, et al. Application of a phage cocktail for control of *Salmonella* in foods and reducing biofilms. Viruses. 2019;**11**(9):841

[128] Meile S, Du J, Dunne M, Kilcher S, Loessner MJ. Engineering therapeutic phages for enhanced antibacterial efficacy. Current Opinion in Virology. 2022;**52**:182-191

[129] Khan MAS, Rahman SR. Use of Phages to treat antimicrobial-resistant *Salmonella* infections in poultry. Veterinary Sciences. 2022;**9**(8):438

[130] Schmelcher M, Loessner MJ. Bacteriophage endolysins: Applications for food safety. Current Opinion in Biotechnology. 2016;**37**:76-87

[131] Kuralkar P, Kuralkar SV. Role of herbal products in animal production—An updated review. Journal of Ethnopharmacology. 2021;**278**:114246

[132] Alagawany M, Elnesr SS, Farag MR, Abd El-Hack ME, Barkat RA, Gabr AA, et al. Potential role of important nutraceuticals in poultry performance and health—A comprehensive review. Research in Veterinary Science. 2021;**137**:9-29

[133] Kotzekidou P, Giannakidis P, Boulamatsis A. Antimicrobial activity of some plant extracts and essential oils against foodborne pathogens in vitro and on the fate of inoculated pathogens in chocolate. LWT-Food Science and Technology. 2008;**41**(1):119-127

[134] Ruvalcaba-Gómez JM, Villagrán Z, Valdez-Alarcón JJ, Martínez-Núñez M, Gomez-Godínez LJ, Ruesga-Gutiérrez E, et al. Non-antibiotics strategies to control *Salmonella* infection in poultry. Animals (Basel). 2022;**12**(1):102

[135] Gibson GR, Hutkins R, Sanders ME, Prescott SL, Reimer RA, Salminen SJ, et al. Expert consensus document: The International Scientific Association for Probiotics and Prebiotics (ISAPP) consensus statement on the definition

and scope of prebiotics. Nature Reviews Gastroenterology & Hepatology. 2017;**14**:491-502

[136] Tran THT, Everaert N, Bindelle J. Review on the effects of potential prebiotics on controlling intestinal enteropathogens *Salmonella* and *Escherichia coli* in pig production. Journal of Animal Physiology and Animal Nutrition. 2018;**102**(1):17-32

[137] Ranjha MMAN, Shafique B, Batool M, et al. Nutritional and health potential of probiotics: A review. Applied Sciences. 2021;**11**(23):11204

[138] Villagran-de la Mora Z, Nuño K, Olga V, Avalos H, Castro-rosas J, Carlos G, et al. Effect of a synbiotic mix on intestinal structural changes, and *Salmonella* Typhimurium and *Clostridium perfringens* colonization in broiler chickens. Animals. 2019;**9**:777

[139] Vinderola G, Sanders ME, Salminen S, Szajewska H. Postbiotics: The concept and their use in healthy populations. Frontiers in Nutrition. 2022;**9**:1002213

[140] Gingerich E, Frana T, Logue CM, Smith DP, Pavlidis HO, Chaney WE. Effect of feeding a postbiotic derived from *Saccharomyces Cerevisiae* fermentation as a preharvest food safety hurdle for reducing *Salmonella* Enteritidis in the ceca of layer pullets. Journal of Food Protection. 2021;**84**(2):275-280

[141] Wang HT, Yu C, Hsieh YH, Chen SW, Chen BJ, Chen CY. Effects of albusin B (a bacteriocin) of *Ruminococcus Albus* 7 expressed by yeast on growth performance and intestinal absorption of broiler chickens-its potential role as an alternative to feed antibiotics. Journal of the Science of Food and Agriculture. 2011;**91**:2338-2343

[142] Wang L, Hu C, Shao L. The antimicrobial activity of nanoparticles: Present situation and prospects for the future. International Journal of Nanomedicine. 2017;**12**:1227-1249

[143] Desin TS, Köster W, Potter AA. *Salmonella* vaccines in poultry: Past, present and future. Expert Review of Vaccines. 2013;**12**(1):87-96

[144] Allafchian A, Vahabi MR, Jalali SAH, Mahdavi SS, Sepahvand S, Farhang HR. Design of green silver nanoparticles mediated by *Ferula ovina* Boiss. Extract with enhanced antibacterial effect. Chemical Physics Letters. 2022;**791**:article 139392

[145] McEwen SA, Collignon PJ. Antimicrobial resistance: A one health perspective. Microbiology Spectrum. 2018;**6**(2). DOI: 10.1128/microbiolspec.arba-0009-2017

[146] Chernov VM, Chernova OA, Mouzykantov AA, Lopukhov LL, Aminov RI. Omics of antimicrobials and antimicrobial resistance. Expert Opinion on Drug Discovery. 2019;**14**(5):455-468

Chapter 5

Recent Development and Application of Phage-Mediated Biocontrol Strategies and Detection against *Salmonella*

Rui Liu and Chenxi Huang

Abstract

Salmonella has emerged as one of the most important indicators of contamination with foodborne pathogens. Thus, it is necessary to develop sensitive and stable methods for rapid detection and control of *Salmonella* to ensure food safety. Typically, bacteriophages, shortened to phages, can not only specifically recognize host bacteria but also lyse the pathogens to produce many progeny phages. When phages were applied to detect *Salmonella*, they could not only recognize live bacteria but also achieve signal amplification to improve detection sensitivity further. Meanwhile, phages can also be applied as antibacterial agents against *Salmonella* in raw materials in agriculture, processing environments, and extend shelf-time for food products. In this chapter, we reviewed the up-to-date research development to use phages as recognition elements and antibacterial agents for rapid detection and biocontrol of *Salmonella* in foodstuff and agriculture-related matrices, dissected the unavoidable challenges, and debated the upcoming prospects.

Keywords: *Salmonella*, multidrug resistance, biocontrol, rapid detection, phage, food safety

1. Introduction

1.1 *Salmonella*

Salmonella is an important pathogenic factor for foodborne diseases that has attracted widespread attention. It was first isolated from the intestines of pigs infected with classical swine fever by American bacteriologists Salmon and Smith in 1885 during the cholera pandemic [1]. Due to its low requirements for nutrients and living environment, *Salmonella* is widely distributed in nature and has numerous serotypes. Currently, over 2700 *Salmonella* serovars have been discovered worldwide according to the Kauffmann-White scheme [2]. Traditionally, phenotypic and genotypic typing methods are important for surveillance and outbreak detection of *Salmonella* infections. Among them, phenotypic typing includes serotyping and phage typing. The phenotype-based approach relied on the surface antigens detected by agglutination reactions between the flagella, capsule, or mucus layer on the surface of their bacterial cells, purified proteins, and specific anti-serum. Specifically, the O (lipopolysaccharide), H (flagellum), and capsule (Vi) antigens are the main types and are often used for characterization. Typically, the O-antigen is related to lipopolysaccharides (LPS) in the cell wall of *Salmonella*, with high stability and high-temperature resistance. The flagella proteins (H-antigens) have poor stability and are not resistant to high temperatures [3]. Genomic methods are comprehensive and high-resolution approaches that aid in identifying the species, subtypes, and potential virulence factors of *Salmonella*. Common genomic techniques include whole genome sequencing (WGS) technology, which sequences the entire genome of *Salmonella*, providing more comprehensive genetic information. This includes all genotypes, SNPs, insertions, deletions, and other variations, which can be used for more accurate strain identification, subtype analysis, and source tracking. Due to its excellent discriminatory ability and relatively low cost, it is often used as a routine first- or second-line typing method for national surveillance of foodborne pathogens [4]. Multilocus sequence analysis (MLST) is one of the important bacterial genome typing methods based on the sequences of five to seven housekeeping genes. It is used for bacterial typing by analyzing sequence variations of multiple core genes. In *Salmonella* MLST analysis, a scheme of seven housekeeping genes is commonly used [5]. Gene chip technology, with each chip capable of accommodating thousands of oligonucleotide probes, allows for the simultaneous detection of expression levels of thousands of *Salmonella* genes or specific sequence variations. This greatly reduces false positives caused by cross-contamination between pathogens and enables efficient and straightforward identification of gene expression, virulence factor expression, and host interactions of *Salmonella* [6]. Transcriptomics, through high-throughput RNA sequencing (RNA-Seq) and other methods, studies the transcriptional structure of genes, revealing gene expression patterns of bacteria under different conditions. It can be employed for the identification and classification of *Salmonella*, offering advantages of high sensitivity, high throughput, and ease of operation [7]. In addition to the above techniques, there are also epigenomics, restriction fragment length polymorphism (RFLP) and amplified fragment length polymorphism (AFLP) methods [8, 9]. These techniques accurately determine the genetic relationships between strains and are of great value in reducing outbreaks of *Salmonella*-related diseases, thus widely applied in investigations and research on *Salmonella*.

Different serovars of *Salmonella* have different host spectrums and virulence factors. The serotypes related to human diseases mainly include *Salmonella Typhi*, *Salmonella Paratyphi A*, *Salmonella Paratyphi B, Salmonella Paratyphi C, Salmonella Anatum*, *Salmonella Choleraesuis* in swine, *Salmonella Typhimurium*, *Salmonella Enteritidis*, and *Salmonella Newport*. All *Salmonella* infections are considered zoonoses, except for *Salmonella Typhi*, *Salmonella Paratyphi A, Salmonella Paratyphi B, Salmonella Paratyphi C,* and *Salmonella Anatum* which are human-specific diseases. *Salmonella Choleraesuis* in swine belongs to *Salmonella* partial tropism, which is adapted to specific animals to a certain extent and can cause sepsis. *Salmonella* is a Gram-negative bacteria found in a wide range of hosts and can be associated with disease not only in humans but also in animals.

1.2 Hazardous risks of *Salmonella*

Salmonella is the most common foodborne pathogen and was listed by the World Health Organization (WHO) as one of the most dangerous foodborne pathogens. It is also one of the four major causes of global diarrheal diseases [10, 11]. Most cases of salmonellosis originate from the consumption of contaminated food, including beef, milk, chicken, eggs, or vegetables. Symptoms of salmonellosis include diarrhea, fever, and abdominal spasms that occur 12 to 72 hours after eating contaminated food, and generally last for 4 to 7 days. Sometimes, individuals infected with *Salmonella* may also develop Reiter syndrome in the later stages, characterized by frequent joint pain, eye irritation, and painful urination. Furthermore, patients may even experience bacteremia and sepsis, which can lead to death.

Salmonella is naturally distributed widely and mainly inhabits the intestines of animals, such as birds, livestock, rodents, reptiles, and humans [12]. Food transmission is the main route of human infection with *Salmonella* [13]. While all serotypes of *Salmonella* can be pathogenic to humans, *Salmonella* mainly originates from one or several animals and is transmitted to food-related supply chains and consumer populations through wastewater and sewage in the environment, resulting in infection [14]. *Salmonellosis* is generally caused by contaminated food of animal origin (mainly meat products, egg products, seafood, poultry, and other fresh products, as well as quick frozen products and dairy products) [15–17]. These foods are rich in nutrients, which are conducive to the proliferation of *Salmonella* and exacerbate pollution levels. In addition, the above raw materials or products may cause cross-contamination with other products during breeding, food transportation, storage, and processing operations, increasing the risk of pathogen spread and infection. In recent years, foodborne outbreaks of salmonellosis have been associated with nonanimal-based food contaminated with feces, including fresh vegetables, fruits, grains, and ready-to-eat foods [18]. *Salmonella* infection can be acquired through contaminated foods regularly found in meat products (accounting for 9.0 and 12.5%, respectively), dairy products (accounting for 26.3 and 19.6%, respectively), fruits and vegetables and their products (accounting for 3.7 and 23.5%, respectively), indicating that *Salmonella* can easily adapt to various environments and survive.

1.3 Strategies to monitor *Salmonella*

In recent years, governments worldwide have established multiple warning prevention and control systems for foodborne pathogens. The World Health Organization (WHO) has listed *Salmonella* as a foodborne pathogen with severe hazards and moderate hazards (WHO-GSS, http://www.who.ch/salmsurv) and has established a global *Salmonella* monitoring network to strengthen monitoring of *Salmonella* pollution [19]. The United States has established three monitoring network systems to dynamically monitor the occurrence and trend of foodborne diseases in the United States, including the Foodborne Diseases Active Surveillance Network (Food Net), the National Molecular Subtyping Network (Pulse Net) and the National Antimicrobial Resistance monitoring system (NARMS) [20]. In Europe, it has established the Enter-Net international surveillance network (Enter-Net). In China, *Salmonella* is an important monitoring indicator for foodborne pathogens in the national food pollutant monitoring system initially established by the Nutrition and Food Safety Institute of the China Center for Disease Control and Prevention.

According to the data from the Chinese national food safety risk surveillance system, 43.0% of food safety issues are attributed to foodborne microorganisms. *Salmonella* is a primary contributor to foodborne illnesses worldwide. In China, 70–80% of foodborne disease outbreaks are attributable to *Salmonella*. Among them, *Salmonella Enteritis* and *Salmonella Typhimurium* are the most common. In 2019, out of the total 139 *Salmonella* samples collected in Xiamen City, *Salmonella Enteritis* and *Salmonella Typhimurium* predominated, constituting 40.3 and 19.4%, respectively [21]. In China, meat and meat products are highly susceptible to contamination by *Salmonella* spp. In the past decade (2008–2019), *Salmonella* has continued to be the leading cause of large foodborne outbreaks in the United States, according to the Foodborne Diseases Active Surveillance Network (FoodNet) reports [22–24]. Based on the data from the Foodborne Diseases Active Surveillance Network (FoodNet) of the Centre for Disease Control (CDC) reported that there are nearly 150,000 cases of foodborne diseases caused by *Salmonella* infection in the United States each year, leading to nearly 40,000 hospitalizations and 688 deaths [25]. It ranks first among the foodborne pathogens that cause foodborne diseases. *Salmonella* is the leading cause of food poisoning in European countries. Because of varying dietary habits, vegetables as well as meat products are important sources of *Salmonella* infection [10, 26]. Currently, food safety issues caused by foodborne pathogens are still frequent. For example, in 2020, the US Centers for Disease Control and Prevention reported a *Salmonella* infection incident caused by onions contaminated with *Salmonella*, resulting in a total of 640 infections with *Salmonella* incidents occurring in 43 states in the USA [27].

1.4 Rapid detection methods against *Salmonella*

The typical standard methods for the identification of *Salmonella* bacteria include conventional culture techniques, immunoassays, nucleic acid-based approaches, biochemical analysis, and biosensing methods [28]. For instance, traditional culture-based methods used for *Salmonella* identification are defined by the international standard ISO6579:2002 [29], and the standard method of China [30]. The steps of traditional methods involve preenrichment, selective enrichment, selective isolation, biochemical screening, and serological confirmation using methods including immunoassays and molecular methods, such as polymerase chain reaction (PCR). The entire process takes 4–7 days to complete, and the results are accurate and reliable, which can intuitively reflect the actual situation of the sample [30]. Typically, conventional cultivation methods are accurate, sensitive, and reproducible, but they are time-consuming, cumbersome, and rely on subjective judgment [31].

To avoid complex cultivation processes to reduce workload, rapid detection strategies are increasingly being studied and applied, such as immunoassays and molecular-based technologies. Immunoassays are based on antigen-antibody reactions, which have the advantages of good specificity and sensitivity but still suffer from long detection times and antibodies with high costs [32]. Molecular techniques are based on the detection of specific nucleic acid sequences, including polymerase chain reaction (PCR) [33], loop-mediated isothermal amplification (LAMP) [34], gene chip technology [35]. These methods have significant advantages in specificity, sensitivity, and rapidity but require high requirements for instrument operators and detection conditions, and they cannot distinguish live and dead bacteria [36]. Furthermore, novel technologies such as biosensing have also been developed and utilized, such as microfluidic technology [37]. It is based on the manipulation of trace fluid systems and is particularly suitable for the analysis of small-sized targets. Biosensors are analytical devices that employ specific biomolecules to recognize the target entity. They could integrate the biorecognition element with signal amplification methods to achieve extremely high sensitivity and effective transducers for signal readouts. They have the advantages of high sensitivity, fast detection, and good reproducibility [38]. Biosensors have been developed for rapid and quantitative detection of *Salmonella* in foods. For example, Liu et al. developed a microfluidic-based biosensor that can detect *Salmonella* within 1 hour, with a detection limit of 300 cells/mL [39]. The sensitivity and specificity of biosensors largely depend on the interference of background microorganisms, sample matrices, and inhibitory substances [40]. Nanotechnology has the advantages of being simple, low-cost, fast, sensitive, and easy to operate, and therefore could be effectively integrated with biosensing methods [41]. Different analysis technologies are compared by their advantages and disadvantages to provide novel strategies for more efficient and practical analysis methods [42, 43]. Up to now, the existing detection methods still could not meet the requirements for on-site and point-of-care testing for viable *Salmonella*, and further improvement is still needed to achieve higher sensitivity and specificity.

2. Application of phages for *Salmonella* biocontrol and detection

2.1 Bacteriophages

Bacteriophages, abbreviated as phages, are viruses that specifically infect bacteria and consist of proteins and nucleic acids. The first reported use of bacteriophages can be traced back to the early twentieth century in 1915. British microbiologist Frederick W. Twort first observed phages when culturing *Staphylococcus aureus* but failed to follow up on his original observations [44]. In 1917, microbiologist Felix d'Herelle first isolated phages and applied them to treat bacterial infections. The word "phage" literally means bacteria eaters and comes from the Greek words "bacteria" and "φαγεῖν (phagein)" [45]. From then on, phages began advocating for the development of a new research field in natural science. Phages play an important role in establishing the central principles of molecular biology, providing support for the development of modern biological technologies and reagents, as well as the discovery of the CRISPR-Cas phage resistance system [46, 47].

Phages can be found in the environment in a wide variety and large quantities that are widespread on Earth. Researchers have isolated phages in various environments [48], such as soil, water, animal intestines, and saliva. Even in some extreme environments, small amounts of phages are present, such as in areas with high temperatures (geothermal hot springs and volcanic vents), cold temperatures, and high osmotic pressures (polar lakes and seawater). Previous studies have found that phages can be isolated almost anywhere bacteria live. The total number of natural phages can reach 10^{30}–10^{32}, which is more than ten times the number of bacteria [49]. It is believed that there are 4.7×10^{30} virions in the ocean (including seawater and marine underground), approximately distributed as 2.6×10^{30} virions in soils [50]. Therefore, phages are highly abundant in nature and could be used as low-cost antimicrobials.

In 1928, antibiotics such as penicillin were discovered and widely applied in clinical trials. Under that condition, phage therapy was abandoned to combat bacteria [51]. Russia approved the introduction of phages to the official pharmacopeia in 2016 [52]. In Europe, phage therapy has been officially permitted to be applied in patients with cystic fibrosis by Eastern European health institutions. However, it is still not able to solve ethical concerns in patients with cystic fibrosis using phage therapy in Western European international regulatory bodies [53]. Phages are viruses that specifically infect bacteria. According to their life cycles of reproduction, they can be divided into lytic or lysogenic phages [54, 55]. To infect the host bacteria, the phage first adsorbs to the receptors on the host cell, then injects the genetic material, and then follows a series of processes, such as lytic or chronic phage pathways. After infecting their host bacteria, lytic phages could effectively use the machinery in the host bacteria to replicate their nucleic acid before finally lysing the bacterial cell and releasing many viral particles to start further bacterial infection. In the lysogenic cycle, phages integrate their genetic elements into the bacterial genome, forming a prophage [56]. Its genome, called a prophage, exists in a free plasmid-like state or is integrated into the bacterial chromosome under pressure conditions such as ultra-violet light and radiation. Under stress conditions, prophages could transfer from the lysogenic state to the lytic cycle before producing the phage progeny. The release of lysogenic phage progeny will cause different effects on the host bacteria, such as the progeny particles of filamentous phages, which are "secreted" and released through the outer membrane of the bacteria, which will not cause bacterial lysis but will slow down the growth of the host bacteria. The reproduction cycle of bacteriophages is shown in **Figure 1**.

Phages have many different forms of existence in nature, such as filamentous, rod-shaped, spherical, and quasi-spherical. The majority of phages are composed of icosahedral head structures (capsids) and tail structures. They can also be classified according to the morphological characteristics of viruses and are currently divided into six categories: *Myoviridae*, *Siphoviridae*, *Podoviridae*, *Microviridae*, *Leviviridae*, and *Inoviridae* [47, 58], with the specific morphological characteristics shown in **Figure 1**. Currently, most of the phages discovered and studied belong to tailed phages, including contractile tails, long and noncontractile tails, and short tails, accounting for 96% of the total number of studies [59]. In recent years, phages

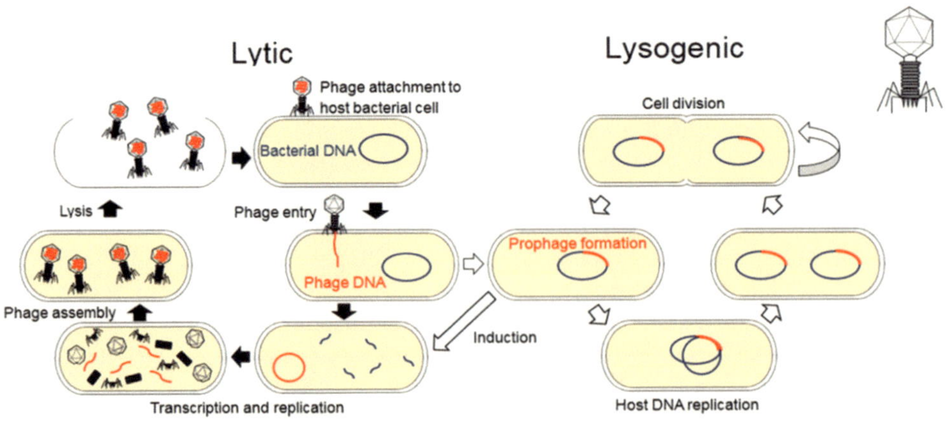

Figure 1.
Phage life cycle [57]. To infect the host bacteria, the phage first adsorbs to the receptors on the host cell, then injects the genetic material, and then follows a series of processes, such as lytic or chronic phage pathways. After infecting their host bacteria, lytic phages could effectively use the machinery in the host bacteria to replicate their nucleic acid before finally lysing the bacterial cell and releasing many viral particles to start further bacterial infection. In the lysogenic cycle, the phage would integrate its genetic elements into the bacterial genome, forming a prophage.

have been recognized as one of the effective alternatives or additions to antibiotics to overcome the challenge of the emergence of antibiotic resistance [60].

2.2 Phage-mediated biocontrol strategies against *Salmonella*

2.2.1 Potential methods for prevention and control of Salmonella

Physical, chemical, antibiotic, and phage methods have been applied to prevent the contamination of *Salmonella* (**Figure 2**). As most of the *Salmonella* strains are sensitive to heat, physical techniques such as ultraviolet radiation and heat treatment have been used to effectively kill or inhibit the growth of *Salmonella* in liquid foods. However, heat treatment can lead to the degradation of nutrients and may also produce advanced glycation end products that are harmful to health [61]. Besides, it may also affect the taste, odor, and texture of foods. For example, Gouma et al. found that when combined heat treatment at 60°C with ultraviolet radiation, the time used for *Salmonella* inactivation in chicken soup can be reduced two times lower than using ultraviolet radiation only. However, the quality of nutrition and sensory texture was reduced as well [62].

Chemical disinfectants have also been widely used for the prevention and control of *Salmonella* [63]. The efficacy of disinfectants was evaluated at a room temperature of 25°C for 5–10 minutes [64]. Results have shown that the disinfection process will require a longer processing time when the temperature decreases [64]. However, none of the tested disinfectants have shown an antibacterial effect on *Salmonella* under dry

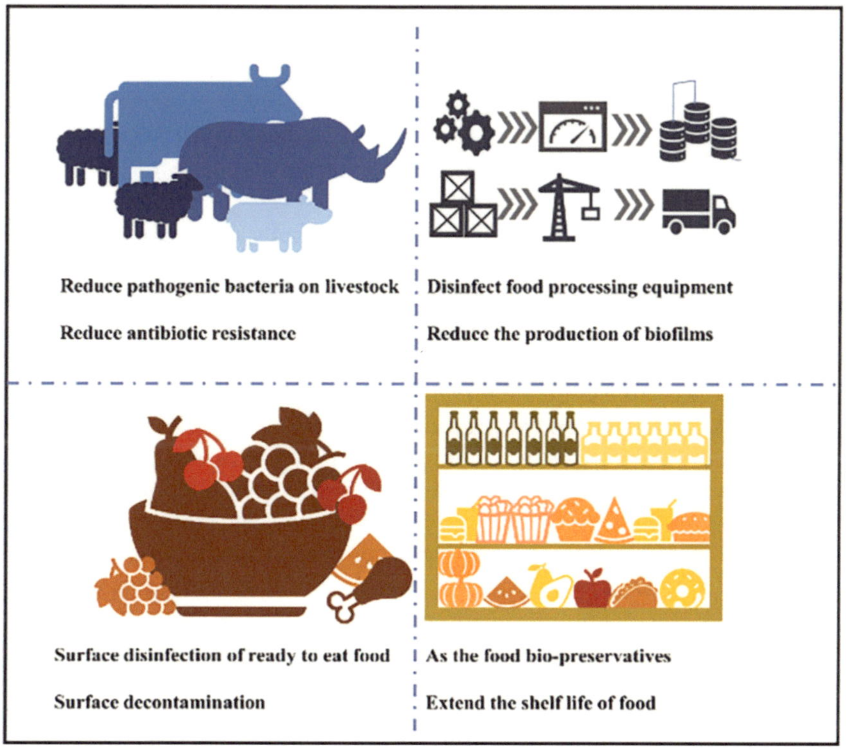

Figure 2.
Phages application in the food industry.

conditions [64]. Disinfectants are non-selective and can rapidly kill a broad spectrum of pathogens. Therefore, it is easy to produce drug-resistant *Salmonella* strains after being exposed to a high concentration of disinfectants frequently. The residues of disinfectants will cause serious issues in the food industry and finally threaten the health of human beings seriously [65]. Previous studies have shown that commonly used disinfectants in poultry processing may confer tolerance to disinfectants and pose a risk of entering the food chain [66].

Antibiotics are a class of chemical substances synthesized by microorganisms that can effectively inhibit or kill bacteria. Antibiotic therapy has been the main strategy to control *Salmonella* infection, which has allowed to improve the growth and health of food-producing animals. Currently, antibiotics used to control *Salmonella* in animal production and human infections mainly include ciprofloxacin, meropenem, and cefepime. They exhibit broad-spectrum antibacterial properties and can eliminate the growth of most pathogens in foods [67]. Antibiotics can inhibit bacterial growth by cross-linking with glycopeptides, binding with phospholipids in cell membranes, or inhibiting bacterial DNA, RNA, and protein synthesis. In recent years, antibiotics have been widely used as feed additives in the field of animal disease prevention during the breeding process. The residues of antibiotics can stably exist in the environment for a long time. The abuse of antibiotics to overcome bacterial infections has caused the emergence of multidrug-resistant bacteria, which pose a great hazard to public health. The mechanisms of antibiotic resistance involve (i) enzyme inactivation, (ii) reduction of intracellular drug accumulation, and (iii) modification of target sites. From 2009 to 2012, studies have reported that 99% of *Salmonella* was found to be antibiotic-resistant in Shandong Province, suggesting that the resistance rate of *Salmonella* has been on the rise for years [68]. The abuse of antibiotics that triggered antibiotic resistance is currently one of the main problems when it comes to *Salmonella* infection. Developing novel strategies to reduce *Salmonella* contamination and prevent biofilm formation in different environments becomes urgently needed.

2.2.2 Phage-mediated biocontrol and prevention strategies against Salmonella

Phages are a type of virus that can infect bacterial cells and replicate and proliferate within bacteria. Phages could effectively kill bacteria with a mechanism that is different from that of antibiotics. Therefore, antibiotic resistance mechanisms do not affect phage efficacy. Phages tend to be natural antibacterial agents with higher safety [69]. Among them, lytic phages can ultimately remove their host bacteria and release progeny phages at the same time. Phages have specific lysis ability on bacteria without affecting the eukaryotic cells. In the presence of sensitive target bacteria, phages can quickly replicate themselves in a short period. Phages are equipped with unique advantages such as high specificity to their hosts, rapid lysis of target bacteria, and the ability to self-proliferate in large quantities [70]. Meanwhile, since there are a large number of phages present in the environment, phage application in the food system can be considered a natural process. More than 10^8 viable bacteriophages per gram may be present in fresh meats and meat products [71]. Phages are considered safe and promising as potential antibacterial agents. However, it is necessary to identify the types of bacteria before utilizing the phages [72].

Phages have been applied in the food industry. Typically, *Salmonella* and other foodborne pathogens that have presented in the process of collection, production, processing, and storage of food raw materials could cause food contamination, leading to food spoilage and foodborne diseases. Phages could be applied as an effective antibacterial agents that can control *Salmonella* in food raw materials, equipment, packaging, and sales to ensure food safety, which includes the following aspects:

1. *Disinfection of animal breeding and raw materials in foods*: During the process of raw materials collection such as animal slaughtering, pathogens can easily contaminate the slaughtered material through blood and feces. As a natural antibacterial agent, phages can reduce the colonization of pathogens on cultured animals. During the time of animal breeding, the pathogens in animals can be killed by oral administration of phages. The inherent characteristics of phages reduce the occurrence of antibiotic resistance and can be used as a disinfectant after slaughter. Therefore, it is possible to meet the demand for raw food materials that are free from pathogens and synthetic chemicals.

2. *Disinfection of food processing and equipment in the food industry*: As for nutrient-rich raw materials, the disinfectant residues in the equipment will accelerate the accumulation of pathogenic bacteria on food processing equipment and biofilms formation, leading to serious public health hazards and economic losses caused by food spoilage. The contamination of pathogens in food processing facilities is often closely related to the formation of biofilms. The resistance of biofilms to chemical/physical cleaning and sanitizing procedures poses a severe threat to human health. It is worth noting that there may be sources of pollution on hands, operating platforms, soaking tanks, industrial facilities, and any surfaces in contact with food during the food processing steps. Under these conditions, phages can disinfect contaminated surfaces, such as working surfaces, floors, walls, and processing equipment [73].

3. *Food additives to extend the shelf life*: Phages can also be directly used in ready-to-eat foods such as vegetables to reduce food contamination. Phages and phage cocktails provide a promising strategy not only for antibiotic-sensitive *Salmonella* but also for antibiotic-resistant *Salmonella* strains. The standard form of phage preparation is phage suspension, which can be directly used or sprayed on the surface of foods after dilution. Spraying and immersion techniques are the two coldest methods for disinfecting phages. Phages can also be prepared into freeze-dried powder or on packaging paper for food preservation. Phage products can free foods from pathogen contamination, as well as extend their shelf life.

2.2.3 Commercialized phage products

Phages are widely distributed with high specific lysis of host bacteria. They have small genome size, and are easy to modify, making them suitable as antibacterial agents for sanitizing food processing environments and detecting specific bacteria in food products. There are commercialized phage products that can be applied in raw food materials, cleaning production environments, and processing equipment to extend food storage time and disinfect fresh fruits and vegetables (**Figure 3**) [74]. The phage preparation named ListShield™ was approved by the US Food and Drug Administration (FDA) for ready-to-eat meat products in 2006 antibacterial food additives. This approval was extended so that phages could also be applied in the fresh/processed fruits and vegetables, and dairy products in 2015 [75]. Moreover, commercial phage products have also been approved for use against *Salmonella*. For instance, FDA has approved the phage product SalmoFresh™ to be directly used in

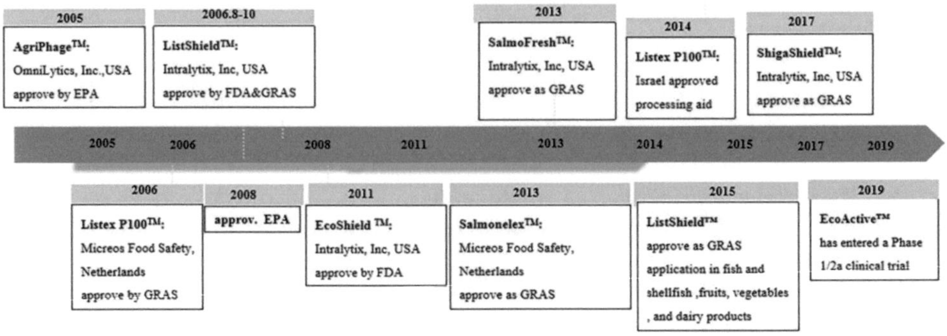

Figure 3.
Approved phage products in the food industry.

meat and poultry, as well as fresh vegetables and fruits [76]. Another phage product Salmonelex™ was approved in 2013 as a processing aid for meat and poultry product production [77]. They have also been applied to fruits and vegetables. For example, Bao et al. [78] mixed two different phages to prepare phage cocktails and sprayed them on the surface of cabbage to inhibit the proliferation of *Salmonella* [78]. The results showed that phage cocktails can effectively inhibit the growth of *Salmonella* at 4°C and 25°C. The effect of phages and chemical disinfectants on *Salmonella* on fruit slices was compared. The antibacterial effect of phages was significantly better than that of chemical disinfectants without affecting the sensory characteristics of fruit quality [79].

2.3 Phage-mediated detection strategies against *Salmonella*

Except for acting as natural antibacterial agents to control foodborne pathogens, phages could also use as recognition elements for the detection of *Salmonella*. Phage receptor binding proteins on its tail could act as receptors for binding to target bacteria hosts. In the presence of target bacteria, the interaction between phage and bacteria could be transferred into measurable detection signals such as optic and electrical signals. Phages were easy to culture and rapidly amplify, thus making their production economical. As a recognition element, phages mainly have unique advantages when it comes to *Salmonella* detection. Firstly, phages are considered highly monodispersed with clear shape and size in a buffer with nanometer range [54]. Secondly, phages can be genetically engineered to display exogenous peptides [80, 81]. Phage particles could also equipped with novel characteristics such as ability to specifically bind to desired target analytes through chemical modification. Thirdly, phages with different characteristics and types were chosen to be used in the biosensors to broaden their detection range. Finally, phages only recognize and infect live bacteria. After characterizing the host range of phages, they can be used for the recognition and detection of target cells [82]. Conversely, phages often have high specificity which could also be problematic, since it would narrow their potential detection spectrum. Multiplexed detection has been developed using different broad-spectrum phages as the probes [83]. Meanwhile, phages can also be used to amplify signals or develop detection methods based on biosensor principles, such as electrochemical and optical biosensors [84]. At present, the main methods for detecting specific pathogens using phages include (i) Traditional phage detection methods (such as phage typing and phage amplification), (ii) phage-labeled bioluminescence and chemiluminescence, (iii) reporter phages construction with molecular biology techniques, (iv) phage-derived components and enzymes, and (v) phages-mediated biosensors [85].

2.3.1 Conventional culture-based detection methods based on phage plaque assay

Lytic phages can specifically recognize and lyse host bacteria to produce progeny phages. The host bacteria were confirmed based on a visual examination of phage plaques on the plates. Traditional phage detection methods determine the quantity of host bacteria based on the number and size of phage plaques, including phage typing and phage amplification strategies. Phage typing is an important bacterial typing method, which is simple, easy to operate, low cost, and rapid. At present, phage typing methods have been developed for the detection of foodborne pathogens such as *Campylobacter*, *Escherichia coli*, *Salmonella*, and *Listeria monocytogenes* [86].

2.3.2 Phage-mediated bioluminescence and chemiluminescence

The bacterial intracellular components, such as adenosine triphosphate (ATP), adenylate kinase (AK), and beta-glucosidase, have been released upon phage infection and could be used as the signal probes for *Salmonella* detection. Firefly luciferase emits light in the presence of ATP and AK. The intensity of light emitted correlates with the amount of ATP, which enables luciferase to catalyze the luminescent reaction. When it integrated with phage amplification assay, it was possible to detect *Staphylococcus aureus*. However, it is insensitive due to the influence of ATP background [87]. In addition, reducing coenzyme amide adenine dinucleotide (NADH) can also be used as a marker to determine the target bacteria under the action of FMN-NADH oxidoreductase and luciferase [88]. The ATP/NADH bioluminescence method is simple and easy to operate. It had a short detection time, high sensitivity, low cost, and no need for cultivation. These methods have been used for the detection of foodborne pathogens such as *Salmonella*, *Escherichia coli*, and *Bacillus* under various conditions. However, this method suffers from the drawbacks of poor stability and requirements of specialized instruments.

Fluorescent labeling detection techniques were developed based on the fluorescent dye for detecting pathogens. The quantum dots (QDs) were prepared as fluorescent labels for rapid and sensitive detection of pathogenic bacteria in food products. A previous study has combined immunomagnetic separation with a quantum dot-labeled phage for the detection of *Salmonella* in milk [89]. The quantum dots (QDs) were prepared as fluorescent labels for rapid and sensitive detection of *Salmonella* in food products, providing a low detection limit (2 CFU/mL) [90]. Fluorescent labeling detection technology has the advantages of high throughput, accuracy, and rapidity. However, it still suffers from high cost and the inability to distinguish between live and dead bacteria.

2.3.3 Phage-mediated biosensing methods

Phage-mediated biosensors have been developed to combine phages with biosensors (**Figure 4**). They are ideal bio-receptors for the construction of highly specific and sensitive probes to construct biosensors. By monitoring the signal changes of certain intracellular substances in bacteria through biosensors, pathogenic bacteria can be detected. For instance, phages were integrated with portable magnetic

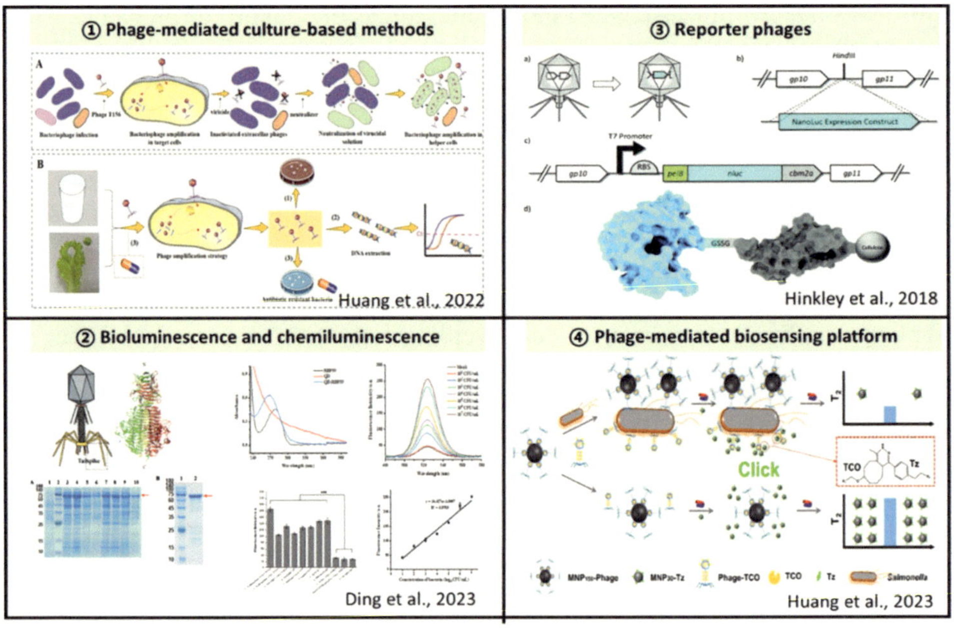

Figure 4.
Phage-mediated detection strategies against Salmonella *[52, 83, 89, 91]. Various phage-mediated methods have been reported in recent years. These methods include culture-based methods, visualized methods using reporter phages, bioluminescence and chemiluminescence, and phage-mediated biosensing platforms.*

relaxation switching biosensors to detect *Salmonella* in orange juice [83]. The method, following a 4-hour preenrichment step, enables the detection of low concentrations of *Salmonella* in orange juice (<10 CFU / 25 mL), with the detected *Salmonella* counts consistent with those obtained using the plate counting method. To improve the sensitivity, the sensing of phage-triggered amplification technology was developed, which utilizes phages to lyse target host bacteria, release intracellular ions, and alter potential potentials [92]. This method has been successfully applied to specifically quantify viable *Salmonella* in lettuce and milk, with a LOD of 1 CFU/mL and the ability to only detect live bacteria. After being coupled with real-time PCR (qPCR), PAA-qPCR further reduced the detection time and achieved an LOD of 10 CFU/mL in 3.5 h without DNA extraction or purification process. The phage-mediated biosensors have the advantages of simple operation and fast speed, but they have shown poor anti-interference ability and low reliability.

2.3.4 Salmonella *detection using reporter phages*

Reporter phages have been defined as genetically modified phages that carry the reporter genes (**Figure 4**). After infection, the reporter gene could be transferred and expressed in the host bacteria upon infection. Expression of reporter genes inside host cells yields a detectable signal, in the presence of target bacteria. Reporter phages have been applied to target the major foodborne pathogens, such as *Salmonella* spp., *Escherichia coli*, *Listeria monocytogenes*, and *Staphylococcus aureus*. Reporter phages containing the luxAB gene were the first bioluminescent for the detection of *Salmonella* strains and were able to detect *Salmonella* in 16 h [93]. In addition, reporter phages P22: inaW, which incorporated the inaW gene of ice nucleation bacteria, were constructed and applied for the detection of *Salmonella* with a detection limit of 10 CFU/mL [94]. This method can detect not only live bacteria but also detect noncultivable bacteria in various environments. Reporter phages-mediated technology was found to have a short time, high sensitivity, and low energy consumption. However, due to the specificity of phages, it is not possible to simultaneously detect all serotypes of the target bacteria.

3. Conclusion

Phages are known to have many advantages, including being able to specifically recognize host bacteria and effectively remove the bacteria. However, there is still uncertainty, especially due to their continuous evolution with host bacteria. Phages may be involved in the transfer of genes encoding toxins. For instance, it has been reported that the first mechanism of phage-mediated gene transfer was identified in *Salmonella* phage P22. High-frequency exchanges of phage genetic materials have enhanced the occurrence of horizontal gene transfer [95]. Naturally, phages and bacteria have been found to transfer an average of approximately 10^{29} genes per day [96]. Therefore, it is necessary to select potent phages through whole genome sequencing instead of temperate phages to control pathogenic bacteria.

Due to the emergence of multidrug-resistant strains, numerous studies have shown that naturally occurring phages can quickly adapt to the bacterial population in their environment. In turn, bacteria can develop resistance to phages [97]. Consequently, phages can combat antibiotic-resistant bacteria, but it is still essential to explore novel strategies to improve phage efficacy, which might be limited by resistance evolution.

Acknowledgements

We thank the Natural Science Foundation of Hubei Province of China (2023AFB330) and the China Postdoctoral Science Foundation (2022 M721275) for financial support.

Conflict of interest

The authors declare no competing financial interest.

Author details

Rui Liu[1,2,3] and Chenxi Huang[1,2,3]*

1 State Key Laboratory of Biocatalysis and Enzyme Engineering, Hubei Key Laboratory of Industrial Biotechnology, School of Life Sciences, Hubei University, Wuhan, China

2 Key Laboratory of Environment Correlative Dietology, Huazhong Agricultural University, Wuhan, Hubei, China

3 College of Food Science and Technology, Huazhong Agricultural University, Wuhan, Hubei, China

*Address all correspondence to: hchenxi543@gmail.com

References

[1] D'Aoust JY. *Salmonella*. In: Doyle MP, editor. Foodborne Bacterial Pathogens. New York: Marcel Inc; 1989. pp. 327-445

[2] Grimont P, Weill FX. Antigenic Formulae of the *Salmonella* Serovars, 9th Edition. WHO Collaborating Centre for Reference and Research on *Salmonella*. World Health Organization; 2007. Available from: https://www.pasteur.fr/sites/default/files/veng_0.pdf

[3] Schatten H, Eisenstark A. *Salmonella*: methods and protocols. In: Methods in Molecular Biology. New Jersey: Humana Press; 2007

[4] Pijnacker R, Dallman TJ, Tijsma AS, Hawkins G, Larkin L, Kotila SM, et al. An international outbreak of *salmonella* enterica serotype Enteritidis linked to eggs from Poland: A microbiological and epidemiological study. The Lancet Infectious Diseases. 2019;**19**(7):778-786

[5] Jiang Z, Li D, Liu Z, Dong X, Liu Z, Cui D, et al. Genomic typing and virulence gene profile analysis of *salmonella* Derby from different sources. Microbial Pathogenesis. 2023;**182**:106248

[6] Courtney S, Mossoba ME, Hammack TS, Keys C, Al-Khaldi SF. Using PCR amplification to increase the confidence level of salmonella typhimurium DNA microarray chip hybridization. Molecular and Cellular Probes. 2006;**20**(3-4):163-171

[7] Zheng L, Shi C, Ma W, Lu Z, Zhou L, Zhang P, et al. Mechanism of biofilm formation by salmonella typhimurium ST19 in a high-glucose environment revealed by transcriptomics. Food Bioscience. 2022;**50**:102074

[8] Bourgeois JS, Anderson CE, Wang L, Modliszewski JL, Chen W, Schott BH, et al. Integration of the salmonella typhimurium methylome and transcriptome reveals that DNA methylation and transcriptional regulation are largely decoupled under virulence-related conditions. MBio. 2022;**13**(3):e03464-e03421

[9] Radomski N, Cadel-Six S, Cherchame E, Felten A, Palma F, Mallet L, et al. A simple and robust statistical method to define genetic relatedness of samples related to outbreaks at the genomic scale–application to retrospective *salmonella* foodborne outbreak investigations. Frontiers in Microbiology. 2019;**10**:483573

[10] Centers for Disease Control and Prevention (CDC). FoodNet fast home page. 2018. Available from: https://wwwn.cdc.gov/foodnetfast/

[11] European Food Safety Authority (EFSA) and European Centre for Disease Prevention and Control (ECDC). The European Union summary report on trends and sources of zoonoses, zoonotic agents, and food-borne outbreaks in 2016. EFSA Journal. 2017;**15**(12):e05077

[12] Liu H, Whitehouse CA, Li B. Presence and persistence of *salmonella* in water: The impact on microbial quality of water and food safety. Frontiers in Public Health. 2018;**6**:159

[13] Gambino D, Gargano V, Butera G, Sciortino S, Pizzo M, Oliveri G, et al. Food is reservoir of MDR *salmonella*: Prevalence of ESBLs profiles and resistance genes in strains isolated from food. Microorganisms. 2022;**10**(4):780

[14] Ran L, Wu S, Gao Y, Zhang X, Feng Z, Wang Z, et al. Foodborne Pathogens and Disease. 2011;**8**(8):921-927

[15] Li Y, Yang X, Zhang H, Jia H, Liu X, Yu B, et al. Prevalence and antimicrobial susceptibility of *salmonella* in the commercial eggs in China. International Journal of Food Microbiology. 2020;**325**:108623

[16] Mamber SW, Mohr T, Barlow K, Bronstein PA, Leathers C, Clinch N. Occurrence of *salmonella* in ready-to-eat meat and poultry product samples from U.S. department of agriculture-regulated producing establishments. II. *Salmonella* in ready-to-eat pork barbecue products, from 2005 to 2012. Journal of Food Protection. 2018;**81**(10):1737-1742

[17] Sonnier JL, Karns JS, Lombard JE, Kopral CA, Haley BJ, Kim SW, et al. Prevalence of *salmonella enterica*, *listeria monocytogenes*, and pathogenic *Escherichia coli* in bulk tank milk and milk filters from US dairy operations in the national animal health monitoring system dairy 2014 study. Journal of Dairy Science. 2018;**101**(3):1943-1956

[18] Paudyal N, Anihouvi V, Hounhouigan J, Matsheka MI, Sekwati-Monang B, Amoa-Awua W, et al. Prevalence of foodborne pathogens in food from selected African countries-a meta-analysis. International Journal of Food Microbiology. 2017;**249**:35-43

[19] World Health Center (WHO). Salmonella-(non-typhoidal). 20 Feb 2018. Available from: https://www.who.int/zh/news-room/fact-sheets/detail/salmonella-(non-typhoidal)

[20] Mutalik VK, Arkin AP. A phage foundry framework to systematically develop viral countermeasures to combat antibiotic-resistant bacterial pathogens. iScience. 2022;**25**(4):104121

[21] Zhang JM, Wen HX, Chen ZH, Cai W, You XW, He SZ. Serotype distribution and genotype analysis of 139 strains of *salmonella* in Xiamen. Modern Preventive Medicine. 2019;**46**:1248-1254

[22] Murray RT, Cruz-Cano R, Nasko D, Blythe D, Ryan P, Boyle M, et al. Prevalence of private drinking water Wells is associated with salmonellosis incidence in Maryland, USA: An ecological analysis using foodborne diseases active surveillance network (FoodNet) data (2007-2016). Science of the Total Environment. 2021;**787**(15):147682

[23] Tack DM, Ray L, Griffin PM, Cieslak PR, Dunn J, Rissman T, et al. Preliminary incidence and trends of infections with pathogens transmitted commonly through food—Foodborne diseases active surveillance network, 10 US sites, 2016-2019. MMWR. Morbidity and Mortality Weekly Report. 2020;**69**(17):509-514

[24] Wang J, Li Y, Xu X, Liang B, Wu F, Yang X, et al. Antimicrobial resistance of *salmonella* enterica Serovar Typhimurium in Shanghai, China. Frontiers in Microbiology. 2017;**8**:510

[25] Habib I, Mohamed MI, Khan M. Current state of *salmonella*, campylobacter and listeria in the food chain across the Arab countries: A descriptive review. Food. 2021;**10**(10):2369. DOI: 10.3390/foods10102369

[26] Tarabees R, Msa E, Shawish R, Basiouni S, Shehata AA. Isolation and characterization of *salmonella* Enteritidis and *salmonella* Typhimurium from chicken meat in Egypt. Journal of Infection in Developing Countries. 2017;**11**:314-319

[27] Centers for Disease Control and Prevention (CDC). Outbreak of

Salmonella Newport infections linked to onions. 2020. Available from: https://www.cdc.gov/salmonella/newport-07-20/index.html

[28] Kyung ML, Mick R, Timothy JH, Robert P, John H. Review of *salmonella* detection and identification methods: Aspects of rapid emergency response and food safety. Food Control. 2015;**47**:264-276

[29] ISO. International Organization for Standardization 6579-2002(E). Microbiology of food and animal feeding stuffs-horizontal method for the detection of *Salmonella* spp. 2002

[30] GB 4789.4-2010. Ministry of Health of China, National food safety standard-Food microbiological examination: *Salmonella*. 2010

[31] Rohde A, Hammerl JA, Boone I, Jansen W, Fohler S, Klein G, et al. Overview of validated alternative methods for the detection of foodborne bacterial pathogens. Trends in Food Science & Technology. 2017;**62**:113-118

[32] Du Z, Wang Y, He D, Xu E, Chai Q, Jin Z, et al. Improving the sensitivity of lateral flow immunoassay for *salmonella* typhimurium detection via flow-rate regulation. Food Chemistry. 2022;**397**:133756

[33] Yang YY, Rajendran V, Jayaraman V, Wang TH, Bei K, Krishna K, et al. Evaluation of the vibrant DNA microarray for the high-throughput multiplex detection of enteric pathogens in clinical samples. Gut Pathogens. 2019;**11**(1):51

[34] Kokkinos PA, Ziros PG, Bellou M, Vantarakis A. Loop-mediated isothermal amplification (LAMP) for the detection of *salmonella* in food. Food Analytical Methods. 2013;**7**:512-526

[35] Sarengaowa HW, Feng K, Jiang AL, Xiu ZL, Lao Y, et al. An in situ-synthesized gene chip for the detection of food-borne pathogens on fresh-cut cantaloupe and lettuce. Frontiers in Microbiology. 2020;**10**:3089

[36] Richter Ł, Janczuk-Richter M, Niedziółka-Jönsson J, Paczesny J, Holyst R. Recent advances in bacteriophage-based methods for bacteria detection. Drug Discovery Today. 2018;**23**:448-455

[37] Wang SY, Zheng LY, Cai GZ, Lin N, Liao M, Li YB, et al. A microfluidic biosensor for online and sensitive detection of *salmonella* Typhimurium using fluorescence labeling and smartphone video processing. Biosensors and Bioelectronics. 2019;**140**:111333

[38] Ahmed A, Rushworth JV, Hirst NA, Millner PA. Biosensors for whole-cell bacterial detection. Clinical Microbiology Reviews. 2014;**27**(3):631-646

[39] Liu J, Jasim I, Shen Z, Zhao L, Dweik M, Zhang S, et al. A microfluidic based biosensor for rapid detection of *salmonella* in food products. PLoS One. 2019;**14**(5):e0216873

[40] Mozola MA. Genetics-based methods for detection of *salmonella* spp. in foods. Journal of AOAC International. 2006;**2**(2):1-4

[41] Pashazadeh P, Mokhtarzadeh A, Hasanzadeh M, Hejazi M, Hashemi M, de la Guardia M. Nano-materials for use in sensing of *salmonella* infections: Recent advances. Biosensors and Bioelectronics. 2017;**87**:1050-1064

[42] Hice SA, Clark KD, Anderson JL, Brehm-Stecher BF. Capture, concentration, and detection of *salmonella* in foods using magnetic ionic liquids and recombinase polymerase

amplification. Analytical Chemistry. 2019;**91**(1):1113-1120

[43] Li J, Liu Q, Wan Y, Wu X, Yang Y, Zhao R, et al. Rapid detection of trace *salmonella* in milk and chicken by immunomagnetic separation in combination with a chemiluminescence microparticle immunoassay. Analytical and Bioanalytical Chemistry. 2019;**411**(23):6067-6080

[44] García P, Martínez B, Obeso JM, Rodríguez A. Bacteriophages and their application in food safety. Letters in Applied Aicrobiology. 2008;**47**(6):479-485

[45] Keen EC. Phage therapy: Concept to cure. Frontiers in Microbiology. 2012;**3**:238

[46] Donohoue PD, Barrangou R, May AP. Advances in industrial biotechnology using CRISPR-Cas systems. Trends in Biotechnology. 2018;**36**:134-146

[47] Salmond GP, Fineran PC. A century of the phage: Past, present and future. Nature Reviews Microbiology. 2015;**13**:777-786

[48] Matsuzaki S, Uchiyama J, Takemurauchiyama I, Daibata M. Perspective: The age of the phage. Nature. 2014;**509**(7498):S9

[49] Bondydenomy J, Qian J, Westra ER, Buckling A, Guttman DS, Davidson AR, et al. Prophages mediate defense against phage infection through diverse mechanisms. ISME Journal. 2016;**10**:2854-2866

[50] Parikka KJ, Le Romancer M, Wauters N, Jacquet S. Deciphering the virus-to-prokaryote ratio (VPR): Insights into virus-host relationships in a variety of ecosystems. Biological Reviews of the Cambridge Philosophical Society. 2017;**92**:1081-1100

[51] Monk AB, Rees CD, Barrow P, Hagens S, Harper DR. Bacteriophage applications: Where are we now? Letters in Applied Microbiology. 2010;**51**(4):363-369

[52] Huang C, Li J, Wang X, Pan H, Wang J, Chen Y. Phage amplification-based technologies for simultaneous quantification of viable *salmonella* in foodstuff and rapid antibiotic susceptibility testing. Food Research International. 2022a;**156**:111279

[53] Rossitto M, Fiscarelli EV, Rosati P. Challenges and promises for planning future clinical research into bacteriophage therapy against *pseudomonas aeruginosa* in cystic fibrosis. An argumentative review. Frontiers in Microbiology. 2018;**9**:775

[54] Sharma S, Chatterjee S, Datta S, Prasad R, Dubey D, Prasad RK, et al. Bacteriophages and its applications: An overview. Folia Microbiologia (Praha). 2017;**62**:17-55

[55] Maurice CF, Bouvier C, De Wit R, Bouvier T. Linking the lytic and lysogenic bacteriophage cycles to environmental conditions, host physiology and their variability in coastal lagoons. Environmental Microbiology. 2013;**15**:2463-2475

[56] Furfaro LL, Chang BJ, Payne MS. Applications for bacteriophage therapy during pregnancy and the perinatal period. Frontiers in Microbiology. 2017;**8**:2660

[57] Batinovic S, Wassef F, Knowler SA, Rice DTF, Stanton CR, Rose J, et al. Bacteriophages in natural and artificial environments. Pathogens. 2019;**8**(3):100

[58] Ofir G, Sorek R. Contemporary phage biology: From classic models to new insights. Cell. 2018;**172**:1260-1270

[59] Simmonds P, Aiewsakun P. Virus classification – Where do you draw the line? Archives of Virology. 2018;**163**(8):2037-2046

[60] Bano S, Hassan N, Rafiq M, Hassan F, Rehman M, Iqbal N, et al. Biofilms as battlefield armor for bacteria against antibiotics: Challenges and combating strategies. Microorganisms. 2023;**11**(10):2595

[61] Uribarri J, Woodruff S, Goodman S, Cai W, Chen X, Pyzik R, et al. Advanced glycation end products in foods and a practical guide to their reduction in the diet. Journal of the American Dietetic Association. 2010;**110**(6):911-916

[62] Gouma M, Gayán E, Raso J, Condón S, Álvarez I. UV-heat treatments for the control of foodborne microbial pathogens in chicken broth. Biomed Research International. 2015;**2015**:436030

[63] Juneja VK, Dwivedi HP, Yan X. Novel natural food antimicrobials. Annual Review of Food Science and Technology. 2012;**3**:381-403

[64] Jang Y, Lee K, Yun S, Lee M, Song J, Chang B, et al. Efficacy evaluation of commercial disinfectants by using *salmonella enterica* serovar Typhimurium as a test organism. Journal of Veterinary Science. 2017;**18**(2):209-216

[65] Pawlowska AM, Zannini E, Coffey A, Arendt EK. "Green preservatives": Combating fungi in the food and feed industry by applying antifungal lactic acid bacteria. Advances in Food and Nutrition Research. 2012;**66**:217-238

[66] Cadena M, Kelman T, Marco ML, Pitesky M. Understanding antimicrobial resistance (AMR) profiles of *salmonella* biofilm and planktonic bacteria challenged with disinfectants commonly used during poultry processing. Food. 2019;**8**(7):275

[67] Beier RC, Anderson PN, Hume ME, Poole TL, Duke SE, Crippen TL, et al. Characterization of *salmonella enterica* isolates from turkeys in commercial processing plants for resistance to antibiotics, disinfectants, and a growth promoter. Foodborne Pathogens and Disease. 2011;**8**(5):593-600

[68] Lai J, Wu C, Wu C, Qi J, Wang Y, Wang H, et al. Serotype distribution and antibiotic resistance of *salmonella* in food-producing animals in Shandong province of China, 2009 and 2012. International Journal of Food Microbiology. 2014;**180**:30-38

[69] McCallin S, Alam Sarker S, Barretto C, Sultana S, Berger B, Huq S, et al. Safety analysis of a Russian phage cocktail: From metagenomic analysis to oral application in healthy human subjects. Virology. 2013;**443**(2):187-196

[70] Flores CO, Meyer JR, Valverde S, Farr L, Weitz JS. Statistical structure of host-phage interactions. National Academy of Sciences of the United States of America. 2011;**108**(28):E288-E297

[71] Goyal SM, Gerba CP, Bitton G. History and basic properties of bacterial viruses. In: Phage Ecology. New York: John Wiley & Sons; 1987. pp. 289-316

[72] Housby JN, Mann NH. Phage therapy. Drug Discovery Today. 2009;**14**:536-540

[73] D'Accolti M, Soffritti I, Mazzacane S, Caselli E. Bacteriophages as a potential 360-degree pathogen control strategy. Microorganisms. 2021;**9**(2):261

[74] Połaska M, Sokołowska B. Bacteriophages a new hope or a huge

problem in the food industry. AIMS Microbiology. 2019;**5**(4):324-346

[75] Jaewoo B, You-Tae K, Sangryeol R, Ju-Hoon L. Biocontrol and rapid detection of food-borne pathogens using bacteriophages and endolysins. Frontiers in Microbiology. 2016;**7**:474

[76] Intralytics USA, Intralytix wins regulatory approval for phage-based food safety product effective against *Salmonella*. 2013. Available from: http//www.intralytix.com/index.php?page=news&id=63

[77] Phageguard USA. FDA-USDA approves SALMONELEX™ as a processing-aid in the production of meat & poultry products. 2013. Available from: http://www.phageguard.com/fda-usda-salmonelexprocessing-aid/

[78] Bao H, Zhang H, Wang R. Isolation and characterization of bacteriophages of Salmonella enterica serovar Pullorum. Poultry Science. 2011;**90**(10):2370-2377

[79] Chaitiemwong N, Hazeleger WC, Beumer RR. Inactivation of *listeria monocytogenes* by disinfectants and bacteriophages in suspension and stainless steel carrier tests. Journal Food Protection. 2014;**77**:2012-2020

[80] Sui Y, Zhu R, Hu W, Zhang W, Zhu HB, Gong M, et al. Phage display screening identifies a prostate specific antigen (PSA)−/lo prostate cancer cell specific peptide to retard castration resistance of prostate cancer. Translational Oncology. 2021;**14**(3):101020

[81] Xu S, Campisi E, Li J, Fischetti VA. Decontamination of *Escherichia coli* O157:H7 on fresh Romaine lettuce using a novel bacteriophage lysin. International Journal of Food Microbiology. 2021;**341**:109068

[82] Peltomaa R, López-Perolio I, Benito-Peña E, Barderas R, Moreno-Bondi MC. Application of bacteriophages in sensor development. Analytical and Bioanalytical Chemistry. 2016;**408**(7):1805-1828

[83] Huang C, Zhao J, Lu R, Wang J, Nugen SR, Chen Y, et al. A phage-based magnetic relaxation switching biosensor using bioorthogonal reaction signal amplification for *salmonella* detection in foods. Food Chemistry. 2023;**400**:134035

[84] Smartt AE, Xu T, Jegier P, Carswell JJ, Blount SA, Sayler GS, et al. Pathogen detection using engineered bacteriophages. Analytical and Bioanalytical Chemistry. 2012;**402**(10):3127-3146

[85] Farooq U, Yang Q, Ullah MW, Wang S. Bacterial biosensing: Recent advances in phage-based bioassays and biosensors. Biosensors and Bioelectronics. 2018;**118**:204-216

[86] Yang X, Wisuthiphaet N, Young GM, Nitin N. Rapid detection of *Escherichia coli* using bacteriophage-induced lysis and image analysis. PLoS One. 2020;**15**(6):e0233853

[87] Li R, Li Z, Huang C, Ding Y, Wang J, Wang X. Phage-based biosensing for rapid and specific detection of *Staphylococcus aureus*. Microorganisms. 2023;**11**(8):2098

[88] Mei C, Wang J, Lin H. Rapid detection of klesiella by bacterial luciferase system conmined with the bacteriophage lysis. Proceeding of International Conference of Natural Produts and Traditional Medicine. 2009;**16-18**:235-239

[89] Ding Y, Zhu W, Huang C, Zhang Y, Wang J, Wang X. Quantum dot-labeled phage-encoded RBP 55 as a fluorescent nanoprobe for sensitive and specific

detection of *salmonella* in food matrices. Food Chemistry. 2023a;**428**:136724

[90] Ding YF, Zhu WJ, Huang CX, Zhang YM, Wang J, Wang XH. Quantum dot-labeled phage-encoded RBP55 as a fluorescent nanoprobe for sensitive and specific detection of *salmonella* in food matrices. Food Chemistry. 2023b;**428**:136724

[91] Hinkley TC, Garing S, Singh S, Le Ny AM, Nichols KP, Peters JE, et al. Reporter bacteriophage T7NLC utilizes a novel NanoLuc: CBM fusion for the ultrasensitive detection of *Escherichia coli* in water. The Analyst. 2018;**143**(17):4074-4082

[92] Huang Y, Wang W, Zhang Z, Gu Y, Huang A, Wang J, et al. Phage products for fighting antimicrobial resistance. Microorganisms. 2022b;**10**(7):1324

[93] Thouand G, Vachou P, Liu S, Dayre M, Griffiths MW. Optimization and validation of a simple method using P22: luxAB bacteriophage forrapid detection of *salmonella enterica* serotypes a, B, and D in poultry samples. Journal of Food Protection. 2008;**71**(2):380-385

[94] Brigati JR, Ripp SA, Johnson CM, Sayler GS. Bacteriophage-based bioluminescent bioreporter for the detection of *Escherichia coli* O157: H7. Journal of Food Protection. 2007;**70**(6):1386-1392

[95] Canchaya C, Fournous G, Brussow H. The impact of prophages on bacterial chromosomes. Molecular Microbiology. 2004;**53**:9-18

[96] Gregory AC, Zayed AA, Conceição-Neto N, Temperton B, Bolduc B, Alberti A, et al. Marine DNA viral macro- and microdiversity from pole to pole. Cell. 2019;**177**(5):1109-1123

[97] Karn SL, Gangwar M, Kumar R, Bhartiya SK, Nath G. Phage therapy: A revolutionary shift in the management of bacterial infections, pioneering new horizons in clinical practice, and reimagining the arsenal against microbial pathogens. Frontiers in Medicine (Lausanne). 19 Oct 2023;**10**:1209782. DOI: 10.3389/fmed.2023.1209782. PMID: 37928478; PMCID: PMC10620811